# Use R!

*Series Editors*
Robert Gentleman    Kurt Hornik    Giovanni Parmigiani

For further volumes:
http://www.springer.com/series/6991

Karline Soetaert
Jeff Cash
Francesca Mazzia

# Solving Differential Equations in R

 Springer

Karline Soetaert
Department Ecosystem Studies
Royal Netherlands Institute for Sea Research
Yerseke
The Netherlands

Jeff Cash
Mathematics
Imperial College
South Kensington Campus
United Kingdom

Francesca Mazzia
Dipartimento di Matematica
University of Bari
Bari
Italy

*Series Editors:*

Robert Gentleman
Program in Computational Biology
Division of Public Health Sciences
Fred Hutchinson Cancer Research Center
1100 Fairview Avenue, N. M2-B876
Seattle, Washington 98109
USA

Giovanni Parmigiani
The Sidney Kimmel Comprehensive
Cancer Center at Johns Hopkins University
550 North Broadway
Baltimore, MD 21205-2011
USA

Kurt Hornik
Department of Statistik and Mathematik
Wirtschaftsuniversität Wien Augasse 2-6
A-1090 Wien
Austria

R-package diffEq to be downloaded from CRAN URL: http://cran.r-project.org/web/
packages/diffEq
In addition R-code of all examples can be downloaded from Extras.Springer.com, also
accessible via Springer.com/978-3-642-28069-6

ISBN 978-3-642-28069-6          ISBN 978-3-642-28070-2 (eBook)
DOI 10.1007/978-3-642-28070-2
Springer Heidelberg New York Dordrecht London

Library of Congress Control Number: 2012939126

Printed on acid-free paper

Springer is part of Springer Science+Business Media (www.springer.com)

*To Carlo, Roslyn and Antonello*

# Preface

Mathematics plays an important role in many scientific and engineering disciplines. This book deals with the numerical solution of differential equations, a very important branch of mathematics. Our aim is to give a practical and theoretical account of how to solve a large variety of differential equations, comprising ordinary differential equations, initial value problems and boundary value problems, differential algebraic equations, partial differential equations and delay differential equations.

The solution of differential equations using R is the main focus of this book. It is therefore intended for the practitioner, the student and the scientist, who wants to know how to use R to solve differential equations.

When writing his famous book, "A Brief History of Time", Stephen Hawking [2] was told by his publisher that every equation he included in the book would cut its sales in half. When writing the current book, we have been mindful of this, and our main desire is to provide the reader with powerful numerical algorithms written in the R programming language for the solution of differential equations rather than considering the theory in any great detail.

However, we also bear in mind the famous statement of Kurt Lewin which is "there is nothing so practical as a good theory". Therefore each chapter that deals with R examples is preceded by a chapter where the theory behind the numerical methods being used is introduced. It has been our goal that non-mathematicians should at least understand the basics of the methods, while obtaining entrance into the relevant literature that provides more mathematical background. We believe that some knowledge of the fundamentals of the underlying algorithms is essential to use the software in an intelligent way, so the principles underlying the various methods should, at least at a basic level, be explained. Moreover, as this book is in the first place about R the discussion of the numerical methods will be skewed to what is actually available in R.

In the sections that deal with the use of R for solving differential equations, we have taken examples from a variety of disciplines, including biology, chemistry, physics, pharmacokinetics. Many are well-known test examples, used frequently in the field of numerical analysis.

# R as a Problem Solving Environment

The choice of using R [8] may be surprising to people regularly involved in solving numerical problems. Powerful numerical methods for the solution of differential equations are typically programmed in e.g. Fortran, C, Java, or Python. Whereas these solution methods are often made freely available, it is unfortunately the case that one needs considerable programming expertise to be able to use them. In contrast, easy-to-use software is often in rather expensive programs, such as MATLAB, Maple or Mathematica. In line with this, most books that give practical information about how to solve differential equations make use of these big three problem solving environments, or of one of the free-of-charge variants.

Although still not often used for solving differential equations, R is also very well suited as a Problem Solving Environment. Apart from the fact that it is open source software, there are obvious advantages in solving differential equations in a software that is strong in visualisation and statistics. Moreover, more and more students are becoming acquainted with the language as its use in universities is growing rapidly, both for teaching and for research. This creates a unique opportunity to introduce these students to the powerful scientific methods which make use of differential equations.

The potential for using R to solve differential equations was initiated by the release of the R package **odesolve** by Woody Setzer, a biologist holding a bachelor's degree in mathematics from EPA, US [10]. Years later, a communication in the R-journal by Thomas Petzoldt, a biologist from the university of Dresden, Germany [5] showed the potential of R for solving initial value problems of ordinary differential equations in the field of ecology. Recently a number of books have applied R in the field of environmental modelling [12, 19]. Building upon this initial effort, Karline Soetaert, the first author of this book, (a biologist) in 2008 joined forces with Woody Setzer and Thomas Petzoldt to make an improved version of odesolve that was able to solve a much greater variety of differential equations. This resulted in the R package **deSolve** [17], which contains most of the integration methods available in R. Most of the solvers implemented in the R package **deSolve** are based on well-established numerical codes, programmed in Fortran. By using well tested, robust, reliable and powerful codes, more emphasis can be put on making the existing codes more versatile. For instance, most codes can now be used to solve delay differential equations, or to simulate events. Also, great care was taken to make a common interface that is (relatively) easy to apply from the user's point of view. A set of methods to solve partial differential equations by the method-of-lines was added to **deSolve**, while another package, **rootSolve** [11], was devised to efficiently solve partial differential equations and boundary value problems using root solving algorithms. Finally, solution methods for boundary value problems were implemented in R package **bvpSolve** [15], as a cooperation between the three authors from this book.

Because all these R packages share one common author (KS), there is a certain degree of consistency in them, which we hope to demonstrate here (see also [16]).

Quite a few other R packages deal with the implementation of differential equations [6, 13], with the solution of special types of differential equations [1, 3, 4, 7], with statistical analysis of their outputs [9, 14, 20], or provide test problems on which the various solvers can be benchmarked [18].

## About the Three Authors

Mathematics is the playground not only for the mathematician and engineer who devise powerful mathematical techniques to solve particular classes of problems, but also for the scientist who applies these methods to real-world problems. Both disciplines meet at the level of software, the actual implementation of these methods in computer code.

The three authors reflect this duality and come from different disciplines. Jeff Cash and Francesca Mazzia are experts in numerical analysis in general and the construction of algorithms for solving differential equations in particular. In contrast Karline Soetaert is a biologist, with an additional degree in computer science, whose interest in these numerical methods is mainly due to the fact that she uses these algorithms for application in the field of the marine sciences. Although she originally wrote her scientific programs mainly in Fortran, since she came acquainted with R in 2007 she now performs nearly all of her scientific work in this programming environment.

**Acknowledgment** Many people have commented on the first versions of this book. We are very thankful for the reviews provided by Filip Meysman, Dick van Oevelen, Tom Cox, Tom van Engeland, Ernst Hairer, Bill Schiesser, Alexander Ostermann, Willem Hundsdorfer, Vincenzo Casulli, Linda Petzold, Felice Iavernaro, Luigi Brugnano, Raymond Spiteri, Luis Randez, Alfredo Bellen, Nicola Guglielmi, Bob Russell, René Lamour, Annamaria Mazzia, and Abdelhameed Nagy.

## References

1. Couture-Beil, A., Schnute, J. T., & Haigh, R. (2010). **PBSddesolve**: *Solver for delay differential equations*. R package version 1.08.11.
2. Hawking, S. (1988). *A brief history of time*. Toronto/New York: Bantam Books. ISBN 0-553-38016-8.
3. Iacus, S. M. (2009). **sde**: *Simulation and inference for stochastic differential equations*. R package version 2.0.10.
4. King, A. A., Ionides, E. L., & Breto, C. M. (2012). **pomp**: *Statistical inference for partially observed Markov processes*. R package version 0.41-3.
5. Petzoldt, T. (2003). R as a simulation platform in ecological modelling. *R News, 3*(3), 8–16.
6. Petzoldt, T., & Rinke, K. (2007). **simecol**: An object-oriented framework for ecological modeling in R. *Journal of Statistical Software, 22*(9), 1–31.
7. Pineda-Krch, M. (2010). **GillespieSSA**: *Gillespie's stochastic simulation algorithm (SSA)*. R package version 0.5-4.

8. R Development Core Team, (2011). *R: A language and environment for statistical computing.* Vienna: R Foundation for Statistical Computing. ISBN 3-900051-07-0.

9. Radivoyevitch, T. (2008). Equilibrium model selection: dTTP induced R1 dimerization. *BMC Systems Biology, 2*, 15.

10. Setzer, R. W. (2001). *The* **odesolve** *package: Solvers for ordinary differential equations.* R package version 0.1-1.

11. Soetaert, K. (2011). **rootSolve***: Nonlinear root finding, equilibrium and steady-state analysis of ordinary differential equations.* R package version 1.6.2.

12. Soetaert, K., & Herman, P. M. J. (2009). *A practical guide to ecological modelling. Using R as a simulation platform.* Dordrecht: Springer. ISBN 978-1-4020-8623-6.

13. Soetaert, K., & Meysman, F. (2012). Reactive transport in aquatic ecosystems: Rapid model prototyping in the open source software R. *Environmental Modelling and Software, 32*, 49–60.

14. Soetaert, K., & Petzoldt, T. (2010). Inverse modelling, sensitivity and monte carlo analysis in R using package **FME**. *Journal of Statistical Software, 33*(3):1–28.

15. Soetaert, K., Cash, J. R., & Mazzia, F. (2011). **bvpSolve***: Solvers for boundary value problems of ordinary differential equations.* R package version 1.2.2.

16. Soetaert, K., Petzoldt, T., & Setzer, R. W. (2010) Solving differential equations in R. *The R Journal, 2*(2):5–15.

17. Soetaert, K., Petzoldt, T., & Setzer, R. W. (2010). Solving differential equations in R: Package **deSolve**. *Journal of Statistical Software, 33*(9):1–25.

18. Soetaert, K., Cash, J. R., & Mazzia, F. (2011). **deTestSet***: Testset for differential equations.* R package version 1.0.

19. Stevens, M. H. H. (2009). *A primer of ecology with R.* Berlin: Springer.

20. Tornoe, C. W., Agerso, H., Jonsson, E. N., Madsen, H., & Nielsen, H. A. (2004). Non-linear mixed-effects pharmacokinetic/pharmacodynamic modelling in **NLME** using differential equations. *Computer Methods and Programs in Biomedicine, 76*, 31–40.

# Contents

# Chapter 1
# Differential Equations

**Abstract** Differential equations (DEs) occur in many branches of science and technology, and there is a real need to solve them both accurately and efficiently. There are relatively few problems for which an analytic solution can be found, so if we want to solve a large class of problems, then we need to resort to numerical calculations. In this chapter we will give a very brief survey of the theory behind DEs and their solution. We introduce concepts such as analytic and numerical methods, the order of differential equations, existence and uniqueness of solutions, implicit and explicit methods. We end with a brief survey of the different types of differential equations that will be dealt with in later chapters of this book.

## 1.1 Basic Theory of Ordinary Differential Equations

Although the material contained in this section is largely of a theoretical nature it is presented at a rather basic level and the reader is advised to at least skim through it.

### 1.1.1 First Order Differential Equations

The general form taken by a first order ordinary differential equation (*ODE*) is

$$y' = f(x, y), \tag{1.1}$$

which may also be written as

$$\frac{dy}{dx} = f(x, y), \tag{1.2}$$

where $f$ is a given function of $x$ and $y$ and $y$ contained in $\Re^m$ is a vector. Here $x$ is called the independent variable and $y = y(x)$ is the dependent variable.

K. Soetaert et al., *Solving Differential Equations in R*, Use R!,
DOI 10.1007/978-3-642-28070-2_1, © Springer-Verlag Berlin Heidelberg 2012

This equation is called *first order* as it contains no higher derivatives than the first. Furthermore, (1.1) is called an *ordinary* differential equation as y depends on one independent variable only.

## 1.1.2   Analytic and Numerical Solutions

A differentiable function $y(x)$ is a solution of (1.1) if for all $x$

$$y'(x) = f(x, y(x)). \tag{1.3}$$

If we suppose that $y(x_0)$ is known, the solution of (1.3), valid in the interval $[x_0, x_1]$, is obtained by integrating both sides of (1.1) with respect to $x$, to give:

$$y(x) - y(x_0) = \int_{x_0}^{x} f(t, y(t))dt, \quad x \in [x_0, x_1]. \tag{1.4}$$

In some cases this integral can be evaluated exactly to give an equation for $y$, and this is called an *analytic* solution. For example, the equation

$$y' = y^2 + 1, \tag{1.5}$$

has as analytic solution

$$y = \tan(x + c). \tag{1.6}$$

Note the free parameter $c$ that occurs in the solution. It has been known for a long time that the solution of a first order equation contains a free parameter and that this solution is uniquely defined if for example we impose an initial condition of the form $y(x_0) = y_0$ and we suppose that the function $f$ satisfies some regularity conditions. This is important and we will return to it later.

Unfortunately, it is true to say that many ordinary differential equations which appear to be quite harmless, in the sense that we could expect them to be easy to solve, cannot be solved analytically, i.e. the solution can not be expressed in terms of known functions. An illuminating example of this is given in [4, p. 4] where it is shown how "small changes" in the problem (1.5) may make it much harder (or impossible) to solve analytically. Indeed, if equation (1.5) is changed "slightly" to

$$y' = y^2 + x, \tag{1.7}$$

then the solution has a very complex structure in terms of Airy functions [4]. In view of this, and the fact that most "real-life" applications consist of complicated systems of equations, it is often necessary to approximate the solution by solving equation (1.1) *numerically* rather than analytically.

Undergraduate mathematics courses often give the impression that most differential equations can be solved analytically, with numerical techniques being developed to deal with those few classes of equations that have no analytic solution. In fact, the opposite is true: while an analytic solution is extremely useful if it does exist, experience shows that most equations of practical interest need to be solved numerically.

### 1.1.3   Higher Order Ordinary Differential Equations

In the previous section, we considered only the first order differential equation (1.1). Ordinary differential equations can include higher order derivatives as well. For example, second order equations of the form:

$$y'' = f(x, y, y'), \tag{1.8}$$

arise in many practical applications.

Normally, in order to deal with the second order equation (1.8), we first convert it to a system of first order equations. This we do by defining an extra dependent variable, which equals the first order derivative of $y$, in the following way:

$$\begin{aligned} y' &= y_1 \\ y_1' &= f(x, y, y_1). \end{aligned} \tag{1.9}$$

Rather than having one differential equation, we now have a system of two differential equations. Defining $Y = (y, y_1)^T$, (1.9) is of the form (1.1), with $Y \in \mathfrak{R}^2$. As we will see later (Sect. 1.1.4) we need to specify two conditions to define the solution uniquely in this second order case.

As a simple example consider a small stone falling through the air from a tower. Gravity produces an acceleration of $g = 9.8 \text{ ms}^{-2}$, while the air exerts a resistive force which is proportional to the velocity ($v$). The differential equation describing this is:

$$v' = g - kv. \tag{1.10}$$

If we are interested in the distance from the top of the tower ($x$), we use the fact that the velocity $v = x'$, and the equation becomes a second order differential equation:

$$x'' = g - kx'. \tag{1.11}$$

Now, in order to solve (1.11), we rewrite it as two first order equations.

$$\begin{aligned} x' &= v \\ v' &= g - kv. \end{aligned} \tag{1.12}$$

This technique carries over to higher order equations as well. If we are faced with the numerical solution of an $n$th order equation, it is often advisable to first reduce it to a system of $n$ first order equations using the obvious extension of the technique described in (1.9) above. Consider for example the "swirling flow III problem" [1, p. 23], which comprises a second order and a fourth order equation describing the flow between two rotating, coaxial disks. The original problem definition

$$\begin{aligned} g'' &= (gf' - fg')/\varepsilon \\ f'''' &= (-ff''' - gg')/\varepsilon, \end{aligned} \tag{1.13}$$

needs one intermediate variable to represent the higher order derivative of $g$, and three to represent the higher order derivatives of $f$. The corresponding set of first order ODEs is:

$$\begin{aligned} g' &= g_1 \\ g_1' &= (gf_1 - fg_1)/\varepsilon \\ f' &= f_1 \\ f_1' &= f_2 \\ f_2' &= f_3 \\ f_3' &= (-ff_3 - gg_1)/\varepsilon. \end{aligned} \tag{1.14}$$

We will solve this problem in Sect. 11.3.

An exception to the rule is the special case where the first derivative $y'$ is absent from (1.8). In such circumstances it is often better to derive special methods for the solution of

$$y'' = f(x, y), \tag{1.15}$$

rather than to introduce the term $y'$ into the definition of the differential equation [3, p. 261].

### 1.1.4  Initial and Boundary Values

We saw in Sect. 1.1 that the integration of the ODE (1.5) introduced an arbitrary constant $c$ into the solution. As long as the ODE is specified only by (1.1), then any value of $c$ will give a valid solution. To select a unique solution, one extra condition is needed and this determines the value of $c$.

Depending on *where* in the integration interval the extra condition is specified we obtain an *initial* or *boundary* value problem. For example, when extending equation (1.5) by introducing the extra condition

$$y(0) = 1, \tag{1.16}$$

then using the general solution (1.6) we obtain $y(0) = \tan(0 + c) = 1$ or $c = \arctan(1) = \pi/4$. Therefore,

$$\begin{aligned} y' &= y^2 + 1 \\ y(0) &= 1, \end{aligned} \tag{1.17}$$

has the unique solution $y = \tan(x + \pi/4)$ providing that we restrict the domain of $x$ suitably. As the extra condition is specified at the initial point of the integration interval, (1.17) is an *initial value problem* (IVP).

The general representation of a first order IVP is:

$$
\begin{aligned}
y' &= f(x, y) \\
y(x_0) &= y_0,
\end{aligned}
\tag{1.18}
$$

where $y$ can be a vector.

In the case of second order equations such as (1.9) it is necessary to prescribe *two* conditions to define $y$ uniquely. In an initial value problem, both these conditions are prescribed at the initial point $(x_0)$. For example we might have:

$$
\begin{aligned}
y'' &= f(x, y, y') \\
y(x_0) &= y_0 \\
y'(x_0) &= y'_0,
\end{aligned}
\tag{1.19}
$$

or, in first order form,

$$
\begin{aligned}
y' &= y_1 \\
y'_1 &= f(x, y, y_1) \\
y(x_0) &= y_0 \\
y_1(x_0) &= y'_0.
\end{aligned}
\tag{1.20}
$$

If instead we prescribe the solution at two different points $x_0$, $x_f$ in the range of integration, we have a *boundary value problem* (BVP). There are several ways in which to specify these boundary conditions, e.g. :

$$
\begin{aligned}
y'' &= f(x, y, y') \\
y(x_0) &= y_0 \\
y(x_f) &= y_f.
\end{aligned}
\tag{1.21}
$$

### 1.1.5  Existence and Uniqueness of Analytic Solutions

An extremely important question concerns the *existence and uniqueness* of solutions of (1.1). This theory is now quite standard and is given for example in [3, Sect. 1.7]. Following the approach of [1] we determine what is required for the IVP solution to exist, be unique and depend continuously on the data, i.e. be well-posed and then ask that the numerical method has similar behaviour.

Basically the main property that we need to ask for if a problem is to be well-posed is that the function $f(x, y)$, appearing in (1.1) should be continuous in a certain region and be Lipschitz continuous [1] with respect to $y$ in that region. An important sufficient condition for Lipschitz continuity is that $f(x, y)$ has bounded partial derivatives $df_i/dy_j$. A nice summary of this theory is found in [1].

As a simple example of an IVP which does not satisfy the conditions for uniqueness consider

$$y' = -\sqrt{1-y^2}, \qquad y(0) = 1. \tag{1.22}$$

This has at least two solutions: $y = 1$, and $y = \cos(x)$. The uniqueness problem occurs because $df/dy$ is unbounded at $x = 0$. However we can also use our intuition to foresee that there may be difficulties with this problem since if we perturb the initial condition to $y(0) = 1 + \varepsilon$ for any positive $\varepsilon$, the solution becomes complex!

The analytic solution of a given second order boundary value problem is rarely possible to obtain (see [3, Sect. 1.3]). Furthermore a proof of the existence and uniqueness of a solution of a given two point boundary value problem is often much harder than for the initial value case and, indeed, boundary value problems are generally much more difficult to solve than initial value problems. We will consider the solution of boundary value problems in detail in Chap. 10.

## 1.2  Numerical Methods

Having briefly outlined some of the basic theory behind the (analytic) solution of ODEs, we now go on to consider some elementary numerical methods. Basically, in their simplest form, numerical methods start by subdividing the domain of the independent variable $x$ into a number of discrete points, $x_0, x_1 = x_0 + h, ...$, and they calculate the approximate values of the dependent variable $y$ and the derivatives of $y$ with respect to $x$ only at these points. These methods are called *finite difference* methods.

Thus, given a series of integration steps $x_0, x_1, \ldots, x_n$, a numerical method constructs a sequence of values $y_0, y_1, \ldots, y_n$, such that

$$y_n \approx y(x_n), \qquad n \geq 0. \tag{1.23}$$

We note the important notation used here namely that $x_n = x_0 + nh$ is a point where the approximate solution will be computed, $y(x_n)$ is the analytic solution at $x_n$ and $y_n$ is the numerical solution obtained at $x_n$.

### 1.2.1  The Euler Method

One of the oldest and most simple numerical methods for solving the initial value problem

$$\begin{aligned} y' &= f(x,y) \\ y(x_0) &= y_0, \end{aligned} \tag{1.24}$$

is due to Euler. This method can be derived in several ways and we start by using a Taylor series approach. Supposing that $f(x,y)$ is analytic in the neighborhood of the

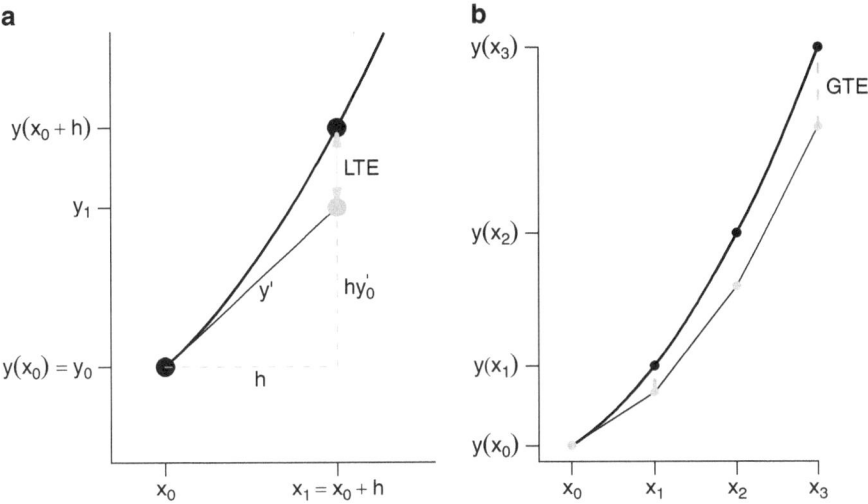

**Fig. 1.1** Errors for the Euler method. (**a**) The local truncation error (*LTE*) is the error introduced by taking one Euler step. (**b**) After taking three integration steps, the global truncation error (*GTE*) is, for sufficiently small $h$, larger than the LTE. This is because for Euler's method the local truncation error is $O(h^2)$ while the global error is $O(h)$

initial value $x_0$, $y_0$ so that we can write

$$y(x_0 + h) = y(x_0) + hy'(x_0) + \sum_{r=2}^{\infty} \frac{h^r}{r!} y^{(r)}(x_0), \qquad (1.25)$$

where $y^{(r)}$ is shorthand for the $r$th derivative of $y$ with respect to $x$. Putting $x_1 = x_0 + h$, ignoring the infinite sum on the right-hand side of (1.25), assuming $y_0$ is exact and denoting the numerical solution obtained at $x_0 + h$ by $y_1$, we obtain the (forward) Euler method:

$$y_1 = y_0 + hf(x_0, y_0). \qquad (1.26)$$

An alternative way of deriving Euler's method is via a geometric approach. If we evaluate the derivative of $y$ at $x_0$ and assume that it is constant throughout $[x_0, x_0 + h]$, we have Fig. 1.1a, which immediately gives $y_1 = y_0 + hy_0'$. Of course $y_0' = f(x_0, y_0)$ so we have again derived Euler's method.

## 1.2.2 Implicit Methods

The Euler formula (1.26) of the previous section expresses the new value $y_1$ as a function of the known $y_0$ and the function value $f(x_0, y_0)$. Thus $y_1$ can be calculated using only known values. Such formulae are called *explicit* methods. When using these methods it is simple to advance to the next integration step.

It is also possible to have *implicit* methods. An example of such a method is the so-called backward Euler method[1]:

$$y_1 = y_0 + hf(x_1, y_1), \tag{1.27}$$

where now the function value $f$ depends on the unknown $y_1$, rather than solely on $y_0$. To solve for $y_1$ is in general not simple: naively we might think that we can just calculate the right-hand side, and estimate $y_1$, but we can do this only if $y_1$ is already known! Usually we will need an iterative method to solve for the unknowns. This means extra work per integration step and finding a solution may not always be easy or even possible (see Sect. 2.6).

### 1.2.3  Accuracy and Convergence of Numerical Methods

An important question concerning equation (1.26) is: how *accurate* is it *locally*? To answer this question, we rewrite (1.25) in the form:

$$y(x_1) - y(x_0) - hf(x_0, y(x_0)) = LTE, \tag{1.28}$$

where $LTE$ is the *local truncation error* introduced by truncating the right-hand side of (1.25) and is given for Euler's method by:

$$LTE = \sum_{r=2}^{\infty} \frac{h^r}{r!} y^{(r)}(x_0). \tag{1.29}$$

Since we do not know the analytic solution $y(x)$ of (1.24), we cannot calculate the local truncation error exactly. The important thing about (1.29) is the power of $h$ in the leading term in the expression for the LTE. For Euler's method this power is 2 and so the LTE is $O(h^2)$ and the method is said to have *accuracy* of order 1. In general, if a method has LTE proportional to $h^{p+1}$, $p \geq 1$, i.e. $|LTE| \leq Ch^{p+1}$ for sufficiently smooth problems and for $h$ sufficiently small then the method is said to be of order $p$. The quantity we are interested in is in general not the LTE, but the *global* error $y_n - y(x_n)$ and for Euler's method this is $O(h)$. Hence Euler's method is said to be *convergent* of order 1 (see Fig. 1.1) and the global error for Euler's method is proportional to the constant step size $h$.

A very similar analysis can be carried out for implicit equations such as (1.27) and it is easy to show that (1.27) is also of order 1.

One strategy to reduce the local truncation error is to reduce the size of the steplength of integration $h$. In general, the higher the order of accuracy of the

---

[1]You may wonder why a formula that uses information "forward" in time is called "backward". This will become clear in Sect. 2.2.3.

numerical method, the more effect such step reduction will have on the LTE. On the other hand, higher order methods require more work for one integration step.

The art in devising a good numerical integration method is to achieve a prescribed accuracy with as little work as possible and this usually means with as few function evaluations as possible. Often this involves changing the step size as we perform the integration. If we use Euler's method the global error is proportional to the maximum step size used.

## 1.2.4 Stability and Conditioning

We complete this introductory chapter with a brief discussion concerning the concepts of *stability* and *conditioning*. The concept of stability is usually applied to initial value problems for differential equations, that of conditioning to boundary value problems. Both concepts relate to the effect small changes in (1.18), either in the function $f$, or in the initial (or boundary) conditions $y(x_0)$, have on the solution. If small changes induce large effects, the problem is said to be unstable or ill-conditioned. Conversely, a problem which has the desirable property that "small changes in the data produce small changes in the solution" is said to be stable or well-conditioned.

We can put the concept of stability on a firm theoretical basis as follows. Consider the initial value problem (1.18). A solution $y(x)$ is said to be stable with respect to the initial conditions $y(x_0)$ if, given any $\varepsilon > 0$, there is a $\delta > 0$ such that any other solution $\hat{y}(x)$ of (1.18) satisfying

$$|y(x_0) - \hat{y}(x_0)| \leq \delta, \tag{1.30}$$

also satisfies

$$|y(x) - \hat{y}(x)| \leq \varepsilon \qquad \text{for all } x > x_0. \tag{1.31}$$

This definition is usually called Lyapunov stability.

### 1.2.4.1 Absolute Stability

When we use a numerical method for the solution of a stable initial value problem it is important to require that the numerical solution has the same behavior as the continuous one. What is done in practice to investigate this is to consider a simple test equation

$$y' = \lambda y, \tag{1.32}$$

which is often called Dahlquist's test equation. If we consider $\lambda$ to be complex with $Re(\lambda) < 0$ we know the true solution of this equation tends to zero as $x \to \infty$. We wish the solution of the numerical method to also behave in this way and if it does we say that $h\lambda$ lies in the stability region of the method. A convenient way to

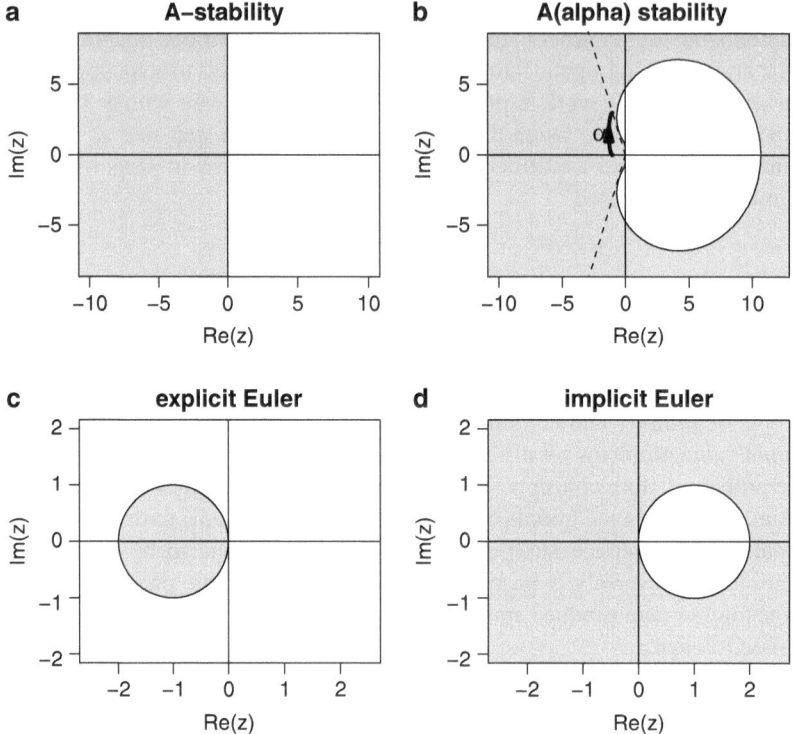

**Fig. 1.2** Stability of numerical methods can be visualised by means of their region of absolute stability (*grey*); of importance here is the region left of the imaginary axis. (**a**) A method is A-stable if its stability region contains at least the left half of the complex plane. (**b**) The absolute stability region for A($\alpha$)-stable methods misses a small part of the left half of the complex plane. (**c**) Only a small part of the left hand complex plane is occupied by the absolute stability region of the explicit Euler method. (**d**) In contrast, the implicit Euler method has a very large region of absolute stability which contains the whole of the complex left half plane

visualise stability properties of a method is to draw its region of absolute stability (see Fig. 1.2). As an example if we consider Euler's method

$$y_1 = y_0 + hf(x_0, y_0), \tag{1.33}$$

applied with a steplength $h$ to $y' = \lambda y$ we obtain

$$y_1 = (1 + z)y_0, \tag{1.34}$$

where $z = h\lambda$. Clearly $|y_1| < |y_0|$ if $|1 + z| < 1$. This region $|1 + z| < 1$ which is the interior of a circle with centre $-1$ and radius $1$ is said to be the region of *absolute stability* or the stability domain of Euler's method.

If instead we consider the backward Euler method we obtain

$$y_1 = \frac{1}{1 - z} y_0. \tag{1.35}$$

Hence the region of absolute stability of the backward Euler method is the exterior of the circle with radius 1 and centre 1. Thus the backward Euler method has a very large region of absolute stability which contains the whole of the complex left half plane (Fig. 1.2d). Such a method is said to be *A-stable*.

A stability requirement that is less severe than A-stability is that of A($\alpha$)-stability. A numerical method is said to be A($\alpha$)-stable ($0 < \alpha < \pi/2$) if its stability region $S \supset S_\alpha = \{\mu : |\arg(-\mu)| < \alpha, \mu \neq 0\}$ (see Fig. 1.2b).

## 1.3   Other Types of Differential Equations

Several other types of differential equations, as well as ordinary differential equations, will be treated in this book.

### 1.3.1   Partial Differential Equations

In the case of ordinary differential equations, the dependent variable is a function of only one independent variable ($x$). In partial differential equations (*PDEs*) there is more than one independent variable.

To distinguish this type of equation from ODEs the derivatives are represented with the $\partial$ symbol rather than with a $d$.

A general representation of a second order PDE with two independent variables, $x_1, x_2$ is:

$$F\left(x_1, x_2, y, \frac{\partial y}{\partial x_1}, \frac{\partial y}{\partial x_2}, \frac{\partial^2 y}{\partial x_1^2}, \frac{\partial^2 y}{\partial x_2^2}, \frac{\partial^2 y}{\partial x_1 \partial x_2}\right) = 0. \tag{1.36}$$

For a partial differential equation to be uniquely specified extra conditions are needed. In each of the independent variables, one needs as many extra conditions as the highest order of the derivatives in this variable.

Similarly to what is done for ODEs, we distinguish between *initial* and *boundary value* problems and this is very important both in the analysis of the problem and in the derivation of numerical methods. When the differential equation contains a combination of initial and boundary conditions it is denoted as an *initial-boundary value* (IBV) problem. For instance, the equation for a *wave* traveling in the $x$-direction is:

$$\frac{\partial^2 u}{\partial t^2} = c^2 \frac{\partial^2 u}{\partial x^2}, \tag{1.37}$$

where $c$ is the phase velocity of the wave, $u$ represents the quantity that changes as the wave passes. Equation 1.37 is a partial differential equation, since $u$ changes both with time $t$ and along the spatial axis $x$. The wave equation is second order in $t$ and $x$. It thus needs two initial conditions and two boundary conditions to be fully specified. This equation will be solved in Sect. 9.2.2.

Another important class of PDEs is the convection-diffusion problem. In one space variable this has the form:

$$\frac{\partial u}{\partial t} = -v\frac{\partial u}{\partial x} + \frac{\partial}{\partial x}\left(q(x)\frac{\partial u}{\partial x}\right). \tag{1.38}$$

Here $q(x) > 0$ is a given function. If it is small in magnitude the PDE is said to be convection dominated. Different numerical methods are often needed to solve (1.38) depending on whether the problem is convection or diffusion dominated.

A powerful way of dealing with time-dependent PDEs is to replace the space derivatives with a finite difference approximation, so producing a system of ODEs to be solved. This technique is called the method of lines and we will return to it in Sect. 8.4.

## 1.3.2  Differential Algebraic Equations

Differential algebraic equations (DAEs) are a mixture of differential and algebraic equations and are represented as:

$$F(x,y,y') = 0. \tag{1.39}$$

As a simple example consider the mathematical pendulum problem in Cartesian coordinates $x, y$:

$$\begin{aligned} mx'' &= -2x\lambda \\ my'' &= -2y\lambda - mg \\ 0 &= x^2 + y^2 - l^2, \end{aligned} \tag{1.40}$$

where $m$ is the pendulum's mass. This DAE comprises a differential part (first two equations) that determines the dynamics of $x$ and $y$ and an algebraic part (third equation) that sets the length of the pendulum equal to $l$. Note that the variable $\lambda$ is not described by a differential equation. The physical meaning of $\lambda$ is the tension in the rod that maintains the mass point on the desired orbit [2].

Insight into how DAEs might be derived from ODEs can be obtained by considering the system of equations

$$\begin{aligned} y' &= f_1(x,y,z) \\ \varepsilon z' &= f_2(x,y,z). \end{aligned} \tag{1.41}$$

If we let $\varepsilon \to 0$ we have in the limit

$$
\begin{aligned}
y' &= f_1(x,y,z) \\
0 &= f_2(x,y,z).
\end{aligned}
\tag{1.42}
$$

This can be regarded as a first order ODE where the variables satisfy algebraic constraints. DAEs are discussed in detail in Chap. 4.

### 1.3.3 Delay Differential Equations

Delay differential equations (*DDEs*) are similar to initial value problems for ODEs, but they involve *past* values of the dependent variables and/or their derivatives. Because of this, rather than needing an initial *value* to be fully specified, DDEs require input of an initial *history* (sequence of values) instead. Thus DDEs are sometimes called differential equations with time lags.

A representation for a DDE which involves past values of the dependent variables only is:

$$
\begin{aligned}
y'(t) &= f(t,y(t),y(t-\tau_1),y(t-\tau_2),...,y(t-\tau_n)) \\
y(t) &= \Phi(t) \qquad \text{for } t < t_0.
\end{aligned}
\tag{1.43}
$$

Here the function $\Phi(t)$ is necessary to provide the history before the start of the simulation (which is at $t_0$).

A very simple DDE example, where the derivative of $y$ is a function of the value of y, lagging with one time-unit $(t-1)$, is:

$$
y' = y(t-1),
\tag{1.44}
$$

for $t > 0$ with history $\Phi(t) = 1$ for $t \le 0$. Note that here $y(t-1)$ denotes the value of $y$ computed at time $t-1$. This type of equation will be dealt with in Chap. 6.

## References

1. Ascher, U. M., Mattheij, R. M. M., & Russell, R. D. (1995). *Numerical solution of boundary value problems for ordinary differential equations*. Philadelphia: SIAM.
2. Hairer, E., & Wanner, G. (1996). *Solving ordinary differential equations II: Stiff and differential-algebraic problems*. Heidelberg: Springer.
3. Hairer, E., Norsett, S. P., & Wanner, G. (2009). *Solving ordinary differential equations I: Nonstiff problems* (2nd rev. ed.). Heidelberg: Springer.
4. Shampine, L. F., Gladwell, I., & Thompson, S. (2003). *Solving ODEs with MATLAB*. Cambridge: Cambridge University Press.

# Chapter 2
# Initial Value Problems

**Abstract** In the previous chapter we derived a simple finite difference method, namely the explicit Euler method, and we indicated how this can be analysed so that we can make statements concerning its stability and order of accuracy. If Euler's method is used with constant time step $h$ then it is convergent with an error of order $O(h)$ for all sufficiently smooth problems. Thus, if we integrate from 0 to 1 with step $h = 10^{-5}$, we will need to perform $10^5$ function evaluations to complete the integration and obtain a solution with error $O(h)$. To achieve extra accuracy using this method we could reduce the step size $h$. This is not in general efficient and in many cases it is preferable to use higher order methods rather than decreasing the step size with a lower order method to obtain higher accuracy. One of the main differences between Euler's and higher order methods is that, whereas Euler's method uses only information involving the value of $y$ and its derivative (slope) at the start of the integration interval to advance to the next integration step, higher order methods use information at more than one point. There exist two important classes of higher order methods that we will describe here, namely Runge-Kutta methods and linear multistep methods.

## 2.1 Runge-Kutta Methods

In this section we will consider the derivation of high order Runge-Kutta methods suitable for the numerical integration of first order systems of ODEs.

### 2.1.1 Explicit Runge-Kutta Formulae

We first consider again the forward Euler method which we have claimed in the previous chapter to be of order 1. Recall that this method is given by:

$$y_{n+1} = y_n + hf(x_n, y_n), \tag{2.1}$$

K. Soetaert et al., *Solving Differential Equations in R*, Use R!,
DOI 10.1007/978-3-642-28070-2_2, © Springer-Verlag Berlin Heidelberg 2012

and its local truncation error is given by:

$$\text{LTE} = y(x_{n+1}) - y_n - hf(x_n, y_n), \tag{2.2}$$

assuming that $y_n \equiv y(x_n)$. To increase accuracy, one approach is to include extra function evaluations in the numerical procedure. Thus, in the case of *Runge-Kutta* methods, the derivatives at several points within the integration interval $[x_n, x_{n+1}]$ are evaluated and used to advance the integration from $x_n$ to $x_{n+1}$.

Runge [34] considered whether it was possible to find high order methods for the numerical solution of (1.1), but it was Kutta [28] who first formulated the general form of a standard explicit Runge-Kutta method as:

$$
\begin{aligned}
y_{n+1} &= y_n + h \sum_{j=1}^{s} b_j k_j \\
k_i &= f\left(x_n + c_i h, y_n + h \sum_{j=1}^{i-1} a_{ij} k_j\right), \quad i = 1, 2, \ldots, s \\
c_i &= \sum_{j=1}^{i-1} a_{ij},
\end{aligned}
\tag{2.3}
$$

where $s$ is the number of intermediate points ("stages") and $k_j$ are the corresponding slopes. This formula gives an approximate solution at $x_{n+1} = x_n + h$, by computing $s$ intermediate stages in the interval $[x_n, x_{n+1}]$.[1]

A great stride forward in the analysis of Runge-Kutta formulae was made by Butcher [9] who suggested writing these formulae in the so-called tableau form:

$$
\begin{array}{c|ccccc}
0 & 0 \\
c_2 & a_{21} & 0 \\
c_3 & a_{31} & a_{32} & 0 \\
\vdots & \vdots & \vdots & \ddots & \ddots \\
c_s & a_{s1} & a_{s2} & \cdots & a_{s,s-1} & 0 \\
\hline
& b_1 & b_2 & \cdots & b_{s-1} & b_s
\end{array}
\tag{2.4}
$$

This can be written in the more concise form

$$
\begin{array}{c|c}
c & A \\
\hline
& b^T
\end{array}
\tag{2.5}
$$

---

[1] Notwithstanding these intermediate stages or steps, you should not be tempted to call a Runge-Kutta method a "multistep" method. This terminology is reserved for a totally different type of method (Sect. 2.2).

Since for the time being we are considering only *explicit* Runge-Kutta methods, the $k_i$ are defined only in terms of $k_1, \ldots, k_{i-1}$, $2 \leq i \leq s$, and consequently the matrix $A$ is strictly lower triangular. Later on we will deal in more depth with *implicit* methods, for which $A$ will have non-zero elements on, and/or above, the diagonal. We now define the *order* of a Runge-Kutta formula [21, p. 133]. Suppose we integrate from $x_0$ to $x_1 = x_0 + h$ using (2.3) and we denote the analytic solution at $x_1$ by $y(x_0 + h)$ and the solution obtained at $x_1$ using (2.3) by $y_1$, then the Runge-Kutta method is said to have order $p$ if, for sufficiently smooth problems (1.1) and $h$ sufficiently small,

$$||y(x_0 + h) - y_1|| < Ch^{p+1}, \tag{2.6}$$

where $C$ is a constant not depending on $h$.

As shown in [21, p. 163], subject to certain smoothness assumptions, if the *local* error is bounded by $Ch^{p+1}$ then the *global* error will be bounded by $Kh^p$ where, similarly to above, $K$ is a constant not depending on $h$.

## 2.1.2  Deriving a Runge-Kutta Formula

In order to be able to derive efficient Runge-Kutta methods several problems, such as high accuracy, stability, provision of continuous solutions, need to be addressed.

### 2.1.2.1  The Order Conditions

One major problem is to determine the coefficients $A$, $b$ and $c$ in (2.5) in such a way that the Runge-Kutta method has a certain order, preferably as high as possible, with as few stages, i.e. function evaluations, as possible. The problem of obtaining the so-called order relations was in the past a bit of a mine field because for example the equations to be solved for are nonlinear and there are more order relations if $y$ is a vector than if $y$ is a scalar. However this was solved by the work of Butcher [9], who derived the order relations for a system. For example, the conditions for Runge-Kutta methods to have order 1, 2 or 3 are respectively:

$$\text{order } 1 : \sum_i b_i = 1 \tag{2.7}$$

$$\text{order } 2 : \sum_i b_i c_i = 1/2 \tag{2.8}$$

$$\text{order } 3 : \sum_i b_i c_i^2 = 1/3 \tag{2.9}$$

$$\sum_{ij} b_i a_{ij} c_j = 1/6, \tag{2.10}$$

where to obtain order 2, we need to satisfy (2.7) and (2.8), while to obtain order 3, we need to satisfy (2.7)–(2.10). Similar conditions exist for higher order methods (see also [21, p. 134] for a fourth order formula). However, the number of order conditions rapidly increases as the required order is increased. For example, to get an order 4 Runge-Kutta formula, eight simultaneous nonlinear algebraic equations need to be solved, while to obtain order 8, we need to solve 200 simultaneous algebraic equations.

### 2.1.2.2  Controlling the Error

Another problem is to control the global error. Unfortunately this is usually not possible so, instead, we estimate and control the local truncation error (LTE). This is done by adapting the step size $h$ in such a way that the estimate of the LTE is smaller than some preset value *tol*. The LTE itself is usually estimated using *embedding*. In this approach two different numerical solutions to the problem, which we denote here by $y_{n+1}$ and $\hat{y}_{n+1}$, are obtained with two different Runge-Kutta formulae, one of order $p + 1$ and one of order $p$ respectively. Assuming that the $O(h^{p+2})$ term in the LTE of $y_{n+1}$ is negligible compared to the $O(h^{p+1})$ term, we will accept the solution $y_{n+1}$ if

$$\|\hat{y}_{n+1} - y_{n+1}\| \leq tol. \tag{2.11}$$

If the error estimate is larger than *tol*, the integration step $h$ will be reduced until the accuracy condition is met [32]. Conversely, if the integration step $h$ produces solutions whose local truncation error is estimated to be much smaller than the preset value, then the solution is accepted and the integration step $h$ is increased.

In practice a mixed absolute-relative error is used based on two tolerances *atol* and *rtol*, and we accept the solution if

$$\|\hat{y}_{n+1} - y_{n+1}\| \leq atol + rtol\|y_{n+1}\|, \tag{2.12}$$

or

$$|\hat{y}_{n+1,i} - y_{n+1,i}| \leq atol_i + rtol_i|y_{n+1,i}|, \tag{2.13}$$

for all $i$, if a componentwise error control is considered. Here $y_{n+1,i}$ denotes the $i$th component of the vector $y_{n+1}$.

Note that this process, known as local extrapolation, is not entirely satisfactory since we attempt to control the error in $\hat{y}_{n+1}$, yet accept the solution $y_{n+1}$ if $\hat{y}_{n+1}$ is sufficiently accurate.

### 2.1.2.3  Changing the Step Size

The way in which we choose the next step length is as follows: suppose we are using a step length $h$ then

$$\|\hat{y}_{n+1} - y_{n+1}\| = h^{p+1}\phi + h.o.t., \tag{2.14}$$

(where $h.o.t.$ means higher order terms in $h$). Suppose we wish to change to a step $\alpha h$. We want

$$(\alpha h)^{p+1}\phi = tol, \tag{2.15}$$

so

$$\alpha = \left(\frac{tol}{\|\hat{y}_{n+1} - y_{n+1}\|}\right)^{1/(p+1)}. \tag{2.16}$$

In practice we allow a margin of error, and use for example

$$\alpha = 0.8\left(\frac{tol}{\|\hat{y}_{n+1} - y_{n+1}\|}\right)^{1/(p+1)}. \tag{2.17}$$

Practical experience shows that this safety factor (0.8) is important since it compensates for us ignoring $h.o.t.$ Similar formulas are used if the mixed absolute-relative error is considered. In this case

$$\alpha = 0.8\left(\frac{atol + rtol\|y_{n+1}\|}{\|\hat{y}_{n+1} - y_{n+1}\|}\right)^{1/(p+1)}, \tag{2.18}$$

is obtained for the normwise error. Similar formulae are obtained if the component-wise error is used.

### 2.1.2.4   Embedded Runge-Kutta Methods

When using the algorithm described above for error estimation the least amount of computational work is needed if the only difference in the two Runge-Kutta methods giving $y_{n+1}$ and $\hat{y}_{n+1}$ (see 2.11) is in the choice of the $b$-coefficients, i.e. we have:

$$\begin{aligned} y_{n+1} &= y_n + h\sum_{j=1}^{s} b_j k_j \\ \hat{y}_{n+1} &= y_n + h\sum_{j=1}^{s} \hat{b}_j k_j, \end{aligned} \tag{2.19}$$

where the first set of coefficients, $b_j$, produces a solution of order $p+1$ and the second set $\hat{b}_j$ a solution of order $p$.

These so-called "embedded Runge-Kutta methods" can be formally represented using the Butcher tableau as:

$$\begin{array}{c|c} c & A \\ \hline & b^T \\ & \hat{b}^T. \end{array} \tag{2.20}$$

Unfortunately, embedded formulae that require no extra function evaluations are only possible for low order Runge-Kutta methods. As the order $p$ increases, extra work must be performed to obtain the embedded solution since additional $k_i$ will need to be computed to allow us to satisfy the order conditions.

Since the order equations are nonlinear it is very difficult, as the order increases, to know how many stages are required to obtain a given order. However Butcher has proved some important order barriers, giving the attainable order for a given number of stages:

$$\begin{array}{c|ccccccccc} \text{Number of stages} & 1\ 2\ 3\ 4\ 5\ 6\ 7\ 8\ 9\ 10 \\ \hline \text{Attainable order} & 1\ 2\ 3\ 4\ 4\ 5\ 6\ 6\ 7\ \ 7 \end{array}$$

This accounts for the great popularity of "the fourth order Runge-Kutta method" since there exists a fourth order method with four stages but there does not exist a fifth order method with five stages and, before the work of Butcher, these high order methods were difficult to derive.

If the order of $y_{n+1}$ and $\hat{y}_{n+1}$ are $q$ and $p$ respectively with $p < q$, we call the Runge-Kutta method a $q(p)$ method. Embedded methods were first introduced in [17] but these have the disadvantage that they give zero error estimates if applied to a quadrature problem. A discussion of embedded formulae is given in [21] and [9] and well-known formulae are the Verner 9(8) pair, the Dormand and Prince 5(4) pair and the Cash-Karp 5(4) pair [9]. The steps taken by the Cash-Karp method are depicted in Fig. 2.1a. The Butcher tableau for the Cash-Karp method is given in the appendix (Sect. A.1).

### 2.1.2.5   Continuous Solutions

Because modern Runge-Kutta methods control an estimate of the local truncation error by adapting the step size $h$, a new problem arises if there is a mismatch between the output positions requested by the user and the values of the independent variable taken by the Runge-Kutta method which will take steps independent of this output "request". Thus, a final problem to be solved for a well-designed Runge-Kutta formula is to provide a solution in between the integration steps, i.e. produce a so-called *continuous solution*. Finding a continuous solution of the same order as the underlying Runge-Kutta formula is not at all straightforward, and consequently not all Runge-Kutta codes have this facility. The interested reader is referred to [21, p. 177] for a discussion of this.

### 2.1.2.6   Stability

In this section we consider the absolute stability region of some explicit Runge-Kutta methods. Recall from Sect. 1.2.4.1 that we compute regions of absolute stability for a given Runge-Kutta method by applying it to the scalar test equation

$$y' = \lambda y. \tag{2.21}$$

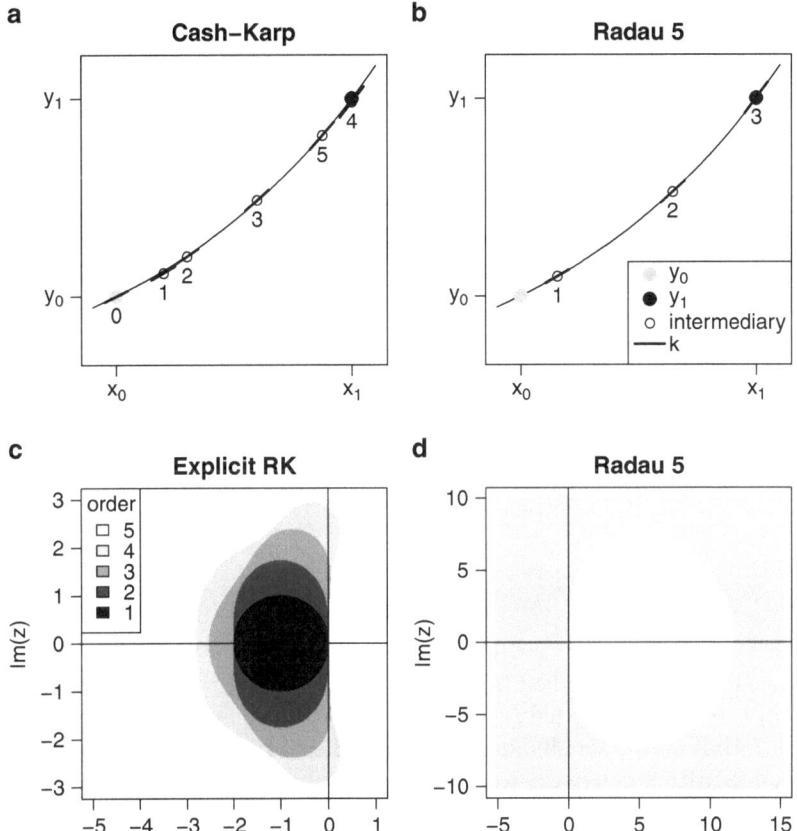

**Fig. 2.1** Properties of Runge-Kutta methods. (**a**) In the (explicit) Cash-Karp method, five inter-mediate stages (1–5) are taken, and the slopes ($k$), at these points evaluated; combined with the known slope at $x_0$, this information is used to advance the integration to $x_1$. (**b**) Although it is of order 5, the implicit Radau method takes only three intermediate steps. This is possible because Radau-5 is implicit while Cash-Karp is explicit. (**c**) The stability regions for explicit Runge-Kutta methods increases with their order, but none of the methods is A-stable. (**d**) In contrast, the stability region for the implicit Radau-5 method comprises the entire region of A-stability (*left* of the imaginary axis); hence Radau-5 is A-stable. However, the Radau Runge-Kutta method is much more expensive than the Cash-Karp method per step

This gives a relationship of the form $y_{n+1} = R(z)y_n$, where

$$R(z) = [1 + zb^T (I - zA)^{-1}\mathbf{1}],\tag{2.22}$$

with $\mathbf{1} = (1, \ldots, 1)^T$. The region of absolute stability of a given Runge-Kutta method is then the set of all $z = h\lambda$ such that $|y_{n+1}| < |y_n|$. If this set includes the whole of the complex left-hand half plane then the method is said to be A-stable. The intersection of $R(z)$ with the negative real axis is called the linear stability boundary.

The absolute stability domains of some well known explicit Runge-Kutta methods are represented in Fig. 2.1c. As can be seen the stability regions do not change very much as the order is increased. A negative stability result is that explicit Runge-Kutta methods cannot be A-stable.

### 2.1.3   Implicit Runge-Kutta Formulae

It is much more straightforward to derive *implicit* Runge-Kutta methods of a certain order than it is to derive explicit Runge-Kutta methods of the same order. Implicit Runge-Kutta methods are of the form

$$y_{n+1} = y_n + h \sum_{j=1}^{s} b_j k_j$$

$$k_i = f(x_n + c_i h, y_n + h \sum_{j=1}^{s} a_{ij} k_j), \quad i = 1, 2, \ldots, s \qquad (2.23)$$

$$c_i = \sum_{j=1}^{s} a_{ij}.$$

We note that for these methods the matrix $A$ is no longer lower triangular. If $a_{ii} \neq 0$ and $a_{ij} = 0$ for $j > i$ we have a *diagonally implicit* Runge-Kutta method (DIRK) [20, p. 91]; if $a_{ii} = \gamma$ for all $i$, the method is called a *singly diagonally implicit* RK method (SDIRK); otherwise, we have an implicit Runge-Kutta method (IRK). As the $k_i$ are defined implicitly, they are computed by solving a set of nonlinear algebraic equations. How to compute this solution efficiently is dealt with in Sect. 2.6.

Getting good step size selection is straightforward for some classes of Runge-Kutta methods, but for others, such as Gauss methods, this is difficult. Also, for Gauss methods it is very difficult to produce continuous output of the same order as the underlying Runge-Kutta formula.

Of particular interest are the Radau IIA methods [20], which are A-stable, implicit Runge-Kutta methods. The intermediate steps taken by the fully implicit Radau IIA method of order 5 are given in Fig. 2.1b, while its stability region is depicted in Fig. 2.1d; the Butcher tableau can be found in the appendix (Sect. A.1).

## 2.2   Linear Multistep methods

An alternative way of obtaining high order accuracy is to derive methods "with memory". Linear multistep methods attempt to approximate $y(x_{n+1})$ as a function of the previously obtained variable values $y_0, y_1, \ldots, y_n$, and/or the previously obtained estimates for the derivatives $f_0, \ldots, f_n$, where $f_i = f(x_i, y_i)$.

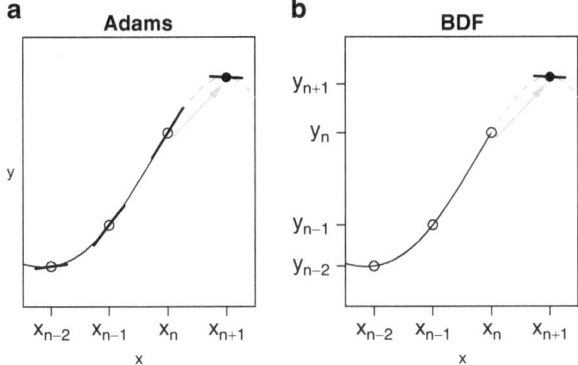

**Fig. 2.2** Schematic representation of the two main classes of linear multistep methods. (**a**) The Adams methods interpolate past values of the slopes, while (**b**) the backward differentiation formulae use past values of $y$

A general linear multistep formula can be written as:

$$\sum_{j=0}^{k} \alpha_{k-j} y_{n+1-j} = h \sum_{j=0}^{k} \beta_{k-j} f(x_{n+1-j}, y_{n+1-j}), \qquad (2.24)$$

where the $\alpha_j$ and $\beta_j$ are real parameters, $h$ is the step size of integration and where $\alpha_k \neq 0$ and $|\alpha_0| + |\beta_0| > 0$. A linear multistep method needs $k-1$ additional starting values, as only $y_0$ is given by the problem. Extra values $y_1, \ldots, y_{k-1}$ should be computed using auxiliary formulae.

If $\beta_k \neq 0$, then the method is implicit, while for $\beta_k = 0$, it is explicit. Note that if $k = 1, \alpha_1 = 1, \alpha_0 = -1, \beta_1 = 1, \beta_0 = 0$ we obtain the backward Euler method, if $\alpha_1 = 1, \alpha_0 = -1, \beta_1 = 0, \beta_0 = 1$ we obtain the forward Euler method. As a consequence, Euler's methods can be regarded both as first order Runge-Kutta methods and as linear multistep methods.

In practice the linear multistep methods that we use have either all the $\beta_i$'s apart from $\beta_k$ zero or else all the $\alpha_i$'s apart from $\alpha_{k-1}$ and $\alpha_k$ zero. The former class of formulae are known as *backward differentiation formulae* (BDF), and they interpolate previous values of $y$. The latter formulae are called *Adams formulae*, and they interpolate the function values $f$ (or slopes) at previous points (see Fig. 2.2).

## 2.2.1   Convergence, Stability and Consistency

In Sect. 1.2.3 the convergence properties of the Euler method were discussed. The generalization to higher order linear multistep methods is however not so straightforward. For a long time, dating back at least to the work of Adams and later from the work of Rutishauser [35], it has been known that high order can give rise to formulae which perform very poorly.

To have a linear multistep method of order $p = 1$ the coefficients should satisfy the following conditions, called conditions of *consistency*:

$$\sum_{j=0}^{k} \alpha_j = 0, \qquad \sum_{j=0}^{k} (j\alpha_j - \beta_j) = 0. \qquad (2.25)$$

More generally conditions for the linear multistep method to have order $p \geq 1$ are (see [21, Theorem III.2.4])

$$\sum_{j=0}^{k} \alpha_j = 0, \qquad \text{and} \qquad \sum_{j=0}^{k} (j^r \alpha_j - r j^{r-1} \beta_j) = 0, \qquad r = 1, 2, \ldots, p. \quad (2.26)$$

The error constant associated with a method of the form (2.24) is defined in, for example, [21, p. 319] and for a $p$th order method it is

$$\frac{1}{(p+1)! \sum_{i=0}^{k} \beta_i} \left( \sum_{j=0}^{k} (j^{p+1} \alpha_j - (p+1) j^p \beta_j) \right). \qquad (2.27)$$

An important property that a linear multistep formula of order $p$ should have in order to have convergence is that it should give a stable solution when applied to the test equation:

$$y' = 0. \qquad (2.28)$$

A linear multistep method applied to (2.28) gives:

$$\sum_{j=0}^{k} \alpha_{k-j} y_{n+1-j} = 0. \qquad (2.29)$$

If we suppose that $y_0, y_1, \ldots, y_{k-1}$ are given starting values, then the solution of this difference equation is stable if the roots of the polynomial

$$\sum_{j=0}^{k} \alpha_j \zeta^j, \qquad (2.30)$$

lie on or inside the unit circle with those on the unit circle being simple. This stability requirement is often referred to as *zero-stability* (see [3]). A fundamentally important result concerning stability, proved by Dahlquist and usually referred to as the *first Dahlquist barrier* is (see [21, p. 384]): the order $p$ of a zero-stable linear $k$-step method satisfies

- $p \leq k + 2$ if $k$ is even
- $p \leq k + 1$ if $k$ is odd
- $p \leq k$ if $\beta_k / \alpha_k \leq 0$.

Zero-stability is an important property because to be convergent a linear multistep method should be zero-stable and consistent. We have the following relation which links convergence, zero-stability and consistency (see [21, Theorem III.4.2], for a proof):

**"Convergence = zero-stability + consistency".**

### 2.2.2   Adams Methods

Consider again the initial value problem

$$\frac{dy}{dx} = f(x,y), \qquad y(x_0) = y_0. \tag{2.31}$$

Suppose now we integrate both sides of the above equation with respect to $x$ to obtain:

$$y(x_{n+1}) = y(x_n) + \int_{x_n}^{x_{n+1}} f(t,y(t))dt. \tag{2.32}$$

On the right-hand side of (2.32) is the unknown function $f(t,y(t))$ integrated over the interval $[x_n, x_{n+1}]$. However, we do know the derivative approximation $f_i = f(x_i, y_i)$ for $n-k+1 \leq i \leq n$. The idea is to replace the function on the right-hand side of (2.32) by the interpolating polynomial $p_k(t)$ of degree $\leq k$, that passes through the known data points $(x_i, f_i, i = n-k+1, n)$. That is, we write:

$$y_{n+1} = y_n + \int_{x_n}^{x_{n+1}} p_k(t)dt, \tag{2.33}$$

where $y_n$ is the previously calculated numerical approximation of $y(x_n)$.

There are several ways in which this interpolating polynomial can be written. In what follows we use backward differences $\nabla^j$, where $\nabla^0 f_n = f_n$, $\nabla^{j+1} f_n = \nabla^j f_n - \nabla^j f_{n-1}$. The interpolating polynomial written in terms of backward differences is

$$p_k(t) = p_k(x_n + sh) = \sum_{j=0}^{k-1} (-1)^j \binom{-s}{j} \nabla^j f_n. \tag{2.34}$$

This is the well-known Newton interpolating polynomial [22] where $\binom{-s}{j}$ is the binomial coefficient. If we now define

$$\gamma_j = (-1)^j \int_0^1 \binom{-s}{j} ds, \tag{2.35}$$

we obtain:

$$y_{n+1} = y_n + h \sum_{j=0}^{k-1} \gamma_j \nabla^j f_n, \tag{2.36}$$

which has order of accuracy equal to $k$ and an error constant $\gamma_k$. These formulae are *explicit* methods, and are called "*Adams-Bashforth*" formulae. Coefficients for these formulae are given in the appendix (Sect. A.2, Table A.3). The simplest method in this class is the forward Euler method.

It is straightforward to define *implicit* Adams methods which we write as:

$$y_{n+1} = y_n + h \sum_{j=0}^{k} \hat{\gamma}_j \nabla^j f_{n+1}, \tag{2.37}$$

where

$$\hat{\gamma}_j = (-1)^j \int_0^1 \binom{-s+1}{j} ds. \tag{2.38}$$

These formulae are implicit since the unknown solution $y_{n+1}$ appears on both sides of (2.37) and the method (2.37) is of order $k+1$. These formulae are called "*Adams-Moulton*" formulae and have as error constant $\hat{\gamma}_{k+1}$. Coefficients for these formulae are given in the appendix (Sect. A.2, Table A.4). The simplest method in this class is the backward Euler method.

A common method of implementing Adams formulae is by using a Predictor-Corrector pair, where a prediction is done using an explicit formula, after which the estimate is corrected using an implicit Adams method. This is discussed in more detail in Sect. 2.6.

### 2.2.2.1 Controlling the Error of Adams Methods

Controlling the local truncation error with Adams methods is straightforward in the case of predictor-corrector methods, since we can design the code so that the difference between the predictor and the corrector gives an estimate of the local error. If for example we use a predictor and a corrector of order $p$ and we assume that the LTE of the predictor and corrector is $\hat{C}_{p+1} h^{p+1} y^{p+1}(x_n)$ and $C_{p+1} h^{p+1} y^{p+1}(x_n)$ respectively then an estimate of the LTE of the corrector is given by $C_{p+1} h^{p+1} y^{p+1}(x_n) \approx \frac{C_{p+1}}{\hat{C}_{p+1} - C_{p+1}} (y_n^C - y_n^P)$ where $y_n^C$ is the corrected value and $y_n^P$ is the predicted value at $x_n$ respectively. This error estimate is normally referred to as Milne's estimate [2, p. 150].

A more commonly used approach is to estimate the local error by comparing a formula of order $p$ with one of order $p+1$. This makes the error easy to estimate when the step size is constant. This is the approach adopted by Shampine and Gordon and is described in [39, p. 26]. However things are much more difficult when using variable step size. A rather general approach has been proposed by Krogh and the interested reader is referred to his original papers and in particular to [27].

#### 2.2.2.2 Stability of Adams Methods

Adams methods are always zero-stable since the polynomial

$$\sum_{j=0}^{k} \alpha_j \zeta^j = \zeta^k - \zeta^{k-1}, \tag{2.39}$$

has one simple root equal to one and the other roots being 0. Suppose we apply an Adams method to the test equation

$$y' = \lambda y. \tag{2.40}$$

In order to have $|y_{n+1}| < |y_n|$ for arbitrary starting values $y_0, \ldots, y_{k-1}$, the roots of the polynomial

$$\zeta^k - \zeta^{k-1} - z \sum_{j=0}^{k} \beta_j \zeta^j, \tag{2.41}$$

with $z = h\lambda$ should lie inside the unit circle.

Some stability regions for Adams methods are shown in Fig. 2.3a, b. It is clear that none of the Adams-Bashforth methods are A-stable. Whereas the first two Adams-Moulton methods (the backward Euler and trapezoidal method) are A-stable, the higher order methods are not (see Fig. 2.3a, b).

### 2.2.3 Backward Differentiation Formulae

One particularly powerful class of methods, proposed by [13], is that of backward differentiation formulae (*BDF*). Recall that the Adams methods were obtained by integrating the polynomial which interpolates past values of the derivative $f$. In contrast, the BDF are derived by differentiating the polynomial which interpolates past values of $y$. The $k$-step BDF which has order $p = k$ is:

$$\sum_{i=1}^{k} \frac{1}{i} \nabla^i y_{n+1} = h f(x_{n+1}, y_{n+1}). \tag{2.42}$$

Writing this in the form of (2.24) with $\alpha_k = 1$ we have:

$$\sum_{i=0}^{k} \alpha_{k-i} y_{n+1-i} = h\beta_k f(x_{n+1}, y_{n+1}). \tag{2.43}$$

Thus, in contrast to the class of Adams formulae, BDF methods have $\beta_k \neq 0$, $\beta_i = 0$ ($i = 0, 1, \ldots, k-1$) and allocate the free parameters to the $\alpha_i$. Coefficients for these

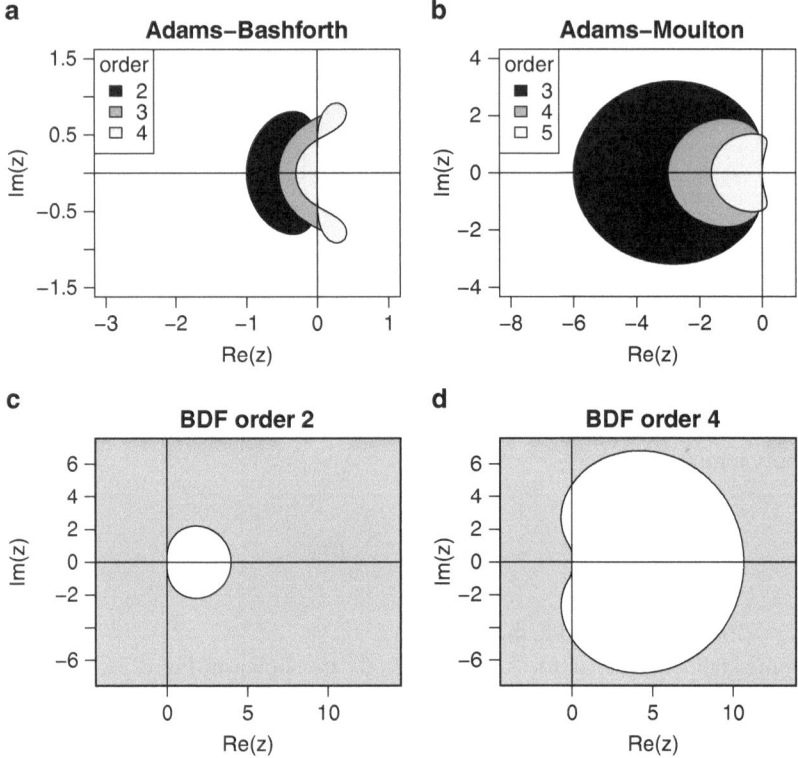

**Fig. 2.3** Stability properties of linear multistep methods for different orders. (**a**) None of the Adams-Bashforth methods are A-stable. (**b**) None of the Adams-Moulton methods of order $\geq$ 3 are A-stable. However, the first two Adams-Moulton methods, the backward Euler (order $= 1$) and the trapezoidal method (order $= 2$) are A-stable and not depicted.(**c**) A-stability is achieved only by BDF methods of order $\leq 2$. (**d**) BDF methods of order 3–6 are A($\alpha$)-stable

formulae are given in the appendix (Sect. A.2, Table A.5). The implicit Euler method can be considered to be the simplest amongst the BDF methods.

### 2.2.3.1    Variable Step Size in BDF Formulae

The way in which the step size of integration is changed when using a linear multistep method is to compute a polynomial $p(x)$ which passes through known data and then to approximate $f(x_{n+1}, y(x_{n+1}))$ by $p'(x_{n+1})$. There are many ways in which we can store the polynomial $p$. Nordsieck proposed storing the so called Nordsieck vector [20, p. 360], so that the polynomial is simple to compute and so that step changing is easy. The Nordsieck vector is

$$\left[ y_n, hy_n', \dots, \frac{h^k}{k!} y_n^{(k)} \right]^T . \tag{2.44}$$

Now, as well as computing the next solution $y_{n+1}$ the history array (2.44) is also updated so that its dependence switches from $y_n$ to $y_{n+1}$. A similar technique has been used for changing the step size using implicit Adams methods. The algorithm that is used [21, p. 347] is rather complex and will not be discussed here.

### 2.2.3.2  Stability of BDF

Practical experience has indicated that a desirable property for a numerical method to possess when integrating stiff problems (see Sect. 2.5) is that of A-stability (see Sect. 2.1.2.6). Unfortunately, the famous Dahlquist's second barrier [14] shows that A-stable linear multistep methods are necessarily of order $p \leq 2$, and if the order is 2 then the error constant is less than or equal to $-1/12$ (see Sect. 1.2.4.1). This makes it pointless to try to find high order A-stable BDF. It is however possible to derive linear multistep methods with A($\alpha$)-stability for order $p > 2$. This is fully described in [21, Chap. III].

Figs. 2.3c, d show A- and A($\alpha$)-stability for BDF methods of order 2 and 4 respectively. It should be noted that BDF are zero-stable only for order $\leq 6$. In practice only methods with order $\leq 5$ are used in general purpose codes, since the order 6 method is only A($\alpha$)-stable with $\alpha < 18°$ and practical experience shows that it is normally more efficient not to include this method.

## 2.2.4  Variable Order – Variable Coefficient Formulae for Linear Multistep Methods

In practice most widely used codes for solving ordinary differential equations change order and/or step size dynamically depending on the local properties of the problem that is being solved.

We have dealt with changing step size earlier; in what follows we describe briefly how variable order can be achieved.

### 2.2.4.1  Variable Order

For a long time variable order was regarded as a difficult problem for multistep methods. However, the definition of Adams or BDF methods in terms of backward differences (2.36), (2.37), and (2.42) highlights the important relationships satisfied by formulae of different order. It is then clear how we might derive a variable order formula since to increase the order of a linear multistep method we just need to add one extra backward difference. Based on this, variable order can be achieved in a very elegant way. There are numerous ways of implementing these formulae, and many of the basic ideas can be found in [39].

### 2.2.4.2  Variable Coefficient Methods

Fixed coefficient methods have the advantage that they are conceptually very simple. Unfortunately they tend to be less stable than *variable coefficient* methods, which we consider next. Assume that at the point $x_n$ we construct the quadratic interpolation polynomial based on unequally spaced data, for the second order formula:

$$p(x) = y_n + (x - x_n)\frac{y_n - y_{n-1}}{h_n} + \frac{(x - x_n)(x - x_{n-1})}{h_n + h_{n-1}}\left(\frac{y_n - y_{n-1}}{h_n} - \frac{y_{n-1} - y_{n-2}}{h_{n-1}}\right).$$
(2.45)

We now differentiate this formula to give $p'(x)$. Evaluating at $x_n$ and simplifying, we obtain the variable step formula:

$$h_n f(x_n, y_n) = y_n - y_{n-1} + \frac{h_n^2}{h_n + h_{n-1}}\left(\frac{y_n - y_{n-1}}{h_n} - \frac{y_{n-1} - y_{n-2}}{h_{n-1}}\right).$$
(2.46)

This form clearly shows the dependence of the method on unequal past data. Note that if we put $h_n = h_{n-1} = h$, we obtain the second order BDF method.

## 2.3  Boundary Value Methods

The linear multistep methods discussed in previous sections are implemented in a step-by-step, *forward-step approach*. This means that given the approximations $y_{n+1-j}, j = 1, \ldots, k$ for some integer $n$ we compute the approximation $y_{n+1}$.

In contrast, boundary value methods (BVMs) [8, 18] integrate initial value problems through "boundary value techniques". One of the aims of this boundary value approach is to circumvent the well known Dahlquist-barriers on convergence and stability which are a direct consequence of the step-by-step application [4].

In order to use a boundary value method we should know in advance the interval in which we would like to compute the solution of (2.31). Let us suppose that the problem is defined in the interval $[a = x_0, b]$. If we define a grid $\pi = [x_0, x_1, \ldots, x_N]$ with $h_i = x_i - x_{i-1}, 1 \leq i \leq N$, $x_0 = a$, $x_N = b$, a $k$-step BVM, using constant step size $h_i = h$, is defined by the following equation:

$$\sum_{j=-k_1}^{k_2} \alpha_{j+k_1} y_{n+j} = h \sum_{j=-k_1}^{k_2} \beta_{j+k_1} f_{n+j}, \quad n = k_1, \ldots, N - k_2.$$
(2.47)

The coefficients $\alpha_0, \ldots, \alpha_k$ and $\beta_0, \ldots, \beta_k$ are the coefficients of the linear multistep formula, $k$ is the number of steps of the linear multistep formula, $k_1$ is the number of initial conditions, $k_2 = k - k_1$ is the number of final conditions. If we choose $k_2 = 0$ we obtain the classical linear multistep formulae. In order to be of practical interest the linear multistep formulas must be associated with $k_1 - 1$ initial and $k_2$ final additional conditions. The first $k_1 - 1$ conditions could be derived, for initial value problems, using a forward approach, the last $k_2$ must be computed by using appropriate discretization schemes. Another way to use the linear multistep formulae in a boundary value approach is to use appropriate discretization schemes for both the initial and the final conditions. This technique, which is efficient for the numerical solution of boundary value problems (see Sect. 10.6.5 for details), has some drawbacks for general initial value problems. The main problem is that the computational effort depends on the number of mesh-points used. A natural strategy to prevent this is to divide the integration interval in subintervals and to use in each subinterval a small fixed number of steps. This technique defines the so called Block Boundary Value Methods that are described in [8, Sect. 11.3–11.5]. Results concerning convergence and stability of block boundary value methods are presented in [1, 25].

Every classical subclass of linear multistep formulae has a corresponding subclass of boundary value methods. In particular we have the generalized backward differentiation formulas (GBDF) and the generalized Adams methods (GAM) [8]. For instance, the generalized Adams methods have the form:

$$y_n - y_{n-1} = h \sum_{j=-k_1}^{k_2} \beta_{j+k_1} f_{n+j}, \; n = k_1, \ldots, N - k_2, \quad (2.48)$$

where $k_1 = (k+1)/2$ for odd $k$ and $k_1 = k/2$ for even $k$. The coefficients of the block-GAMs of order 2–5, using constant step size are in [25]. For these orders the resulting methods are always A-stable.

## 2.4 Modified Extended Backward Differentiation Formulae

A particularly powerful class of high order A-stable methods are the *modified extended backward differentiation* formulae (MEBDF) [11]. These formulae are not Runge-Kutta or linear multistep methods but fit into the class of general linear methods which have been investigated by Butcher [9]. We will explain how these methods are implemented by considering the derivation of a formula of order 2. MEBDF require four steps in their implementation:

- Step 1: Predict a solution $\bar{y}_{n+1}$ at $x_{n+1}$ using backward Euler.

$$\bar{y}_{n+1} = y_n + h\bar{y}'_{n+1}$$

- Step 2: Predict a solution $\bar{y}_{n+2}$ at $x_{n+2}$ using backward Euler.

$$\bar{y}_{n+2} = \bar{y}_{n+1} + h\bar{y}'_{n+2}$$

- Step 3: Compute the derivative at $(x_{n+2}, \bar{y}_{n+2})$

$$\bar{f}_{n+2} = f(x_{n+2}, \bar{y}_{n+2})$$

- Step 4: Solve for the solution $y_{n+1}$ at $x_{n+1}$ using the equation

$$y_{n+1} = y_n - h/2(\bar{f}_{n+2} - \bar{f}_{n+1}) + hy'_{n+1}.$$

There is a lot of computational efficiency in this approach since during the steps (1, 2, 4) three systems of algebraic equations are solved and the method is implemented so that these all have the same coefficient matrix. It is easy to show [20, p. 267], [10] that this formula is of order 2 and is A-stable. Because MEBDF are "super implicit", this allows us to achieve A-stability up to order 4 and $A(\alpha)$-stability up to order 9, at a reasonable computational cost. A short summary is given in [20, p. 267].

A very similar approach to that used by BDF to change the step size is used by MEBDF. In MEBDF the natural history array to store is based on differences of y:

$$[f_n, y_n, \nabla y_n, \nabla^2 y_n, \ldots, \nabla^k y_n]^T. \tag{2.49}$$

Again the idea is to systematically update the history array as the solution is calculated. As a result of backward differences being stored the history array will be easily updated.

Coefficients for these MEBDF formulae are given in the appendix (Sect. A.2, Table A.6).

## 2.5   Stiff Problems

In order to choose the most efficient scheme for computing a numerical solution of (1.1), information about the behavior of the solution is required. Important information concerns the "stiffness" of the problem. Dahlquist classified stiff problems as those that have "processes in the system (described by the differential equation) with significantly different time scales" [5, p. 349]. Let us consider the following example [5]:

$$y'(x) = 100(\sin(x) - y(x)), \quad 0 \le x \le b, \quad y(0) = 0 \tag{2.50}$$

where the solution is $y(x) = (\sin(x) - 0.01\cos(x) + 0.01e^{(-100x)})/1.001$. If $b \approx 10^{-2}$ the solution is computed efficiently by any numerical scheme, but if $b \approx 3$ then the problem is stiff and numerical methods with a small absolute stability region may require very small step sizes for the solution in the entire integration interval. This means that explicit methods encounter such severe difficulties that they fail or become extremely inefficient when applied to stiff problems, whereas implicit methods may give a solution with a reasonable (i.e. not too small) size of the step length of integration.

Because of this, one of the major problems we are faced with when numerically integrating a first order system of initial value problems is to determine whether or not the problem is stiff. Ideally if the problem is stiff then we would hope to use a "stiff integrator" such as BDF and if the problem is not stiff we would like to use a "non-stiff integrator" such as Adams or explicit Runge-Kutta formulae.

To make things more difficult, some problems change from being stiff to being non-stiff, or vice versa, over the integration interval. When a problem changes from being non-stiff to being stiff, it can only be solved efficiently using an implicit method, but it may prove very inefficient if the implicit method is used in the non-stiff part as well.

### 2.5.1   Stiffness Detection

The problem of stiffness detection has been considered by Shampine [36–38] and by Skelboe [42] amongst others. Shampine and Gordon [39, p. 132] describe a procedure for testing whether a code based on Adams predictor-corrector formulas is doing too much work on account of the fact that the problem is stiff. In a variable order/variable step code, low order methods will tend to be used when stiffness is present. So, if many consecutive steps of low order are taken, the problem is probably stiff.

An important property that we can use is that when a stiff problem is solved by an explicit code the product of the step size with the dominant eigenvalue $(\bar{\lambda})$ of the Jacobian $J(= \partial f/\partial y)$ lies close to the border of the stability domain. Hairer et al. [20, p. 21] show two ways how this might be used to detect stiffness with an explicit Runge-Kutta formula. The first method, originally proposed by Shampine [40] is based on comparing two error estimates of different orders $err = O(h^p), \widetilde{err} = O(h^q)$, with $q < p$. Usually $err \ll \widetilde{err}$, if the step size is limited by accuracy requirements and $err > \widetilde{err}$ when the step size is limited by stability requirements. The precise way in which the two error estimators are derived is described in [21, p. 21].

The second way is to approximate the dominant eigenvalue of the Jacobian as follows. Let $v$ denote an approximation to the eigenvector corresponding to the dominant eigenvalue of the Jacobian. If the norm of $v$ $(\|v\|)$ is sufficiently small then the Mean Value Theorem tells us that a good approximation to the dominant eigenvalue is

$$|\bar{\lambda}| = \frac{\|f(t,y+v) - f(t,y)\|}{\|v\|}. \tag{2.51}$$

The cost of this procedure is at most 2, but often 1, function evaluations per step. The product $h|\bar{\lambda}|$ can then be compared to the linear stability boundary of the method.

## 2.5.2   Non-stiffness Test

This test considers the possibility of moving from a stiff method to a non-stiff one. Most methods that have been derived to do this are based on a norm bound $\|J\|$ to estimate the dominant eigenvalue of the Jacobian $J$, i.e. the code computes $\|J\|_p$ where normally $p = 1$ or $p = \infty$. The cost of doing this is $O(m^2)$ multiplications, where $m$ is the dimension of the system. This is small compared with the work required to actually perform the numerical integration. However, as is well known, norm bounds tend to overestimate the dominant eigenvalue and this overestimation becomes worse as the dimension of the problem increases. However norm bounds can be tight, and therefore useful, for sparse or banded matrices of large dimension.

## 2.6   Implementing Implicit Methods

It is clear from previous sections that implicit methods can have much better stability properties than explicit methods. But as they introduce non-trivial computational overhead, they have to be carefully implemented.

The overhead is due to the fact that, as the unknown quantity is present both on the right- and left-hand side of the implicit equation, we need to obtain the solution by solving a system of algebraic equations and normally this system is nonlinear.

There are three important techniques that have been used in the implementation of implicit methods. For the purpose of illustration, we will explain them using the simplest implicit method, namely the backward Euler method:

$$y_{n+1} = y_n + hf(x_{n+1}, y_{n+1}). \tag{2.52}$$

## 2.6.1   Fixed-Point Iteration to Convergence

The most straightforward method is to solve the nonlinear system by successive iterations:

$$y_{n+1}^{v+1} = y_n + hf(x_{n+1}, y_{n+1}^v), \qquad v = 0, 1, \ldots \tag{2.53}$$

These iterations are repeated until the difference between successive steps is sufficiently small. This procedure is called fixed-point or functional iteration.

Each iteration is relatively inexpensive, but as convergence is often slow, in general a lot of iterations will need to be performed. Moreover, for stiff problems, the iteration scheme will not converge unless we place severe restrictions on the steplength $h$, and so this scheme is of no practical use for such problems (see for example [29]).

To initiate the method, we can either take $y_{n+1}^0 = y_n$, or use an explicit method to obtain the first guess of $y_{n+1}$; the latter is referred to as a predictor and the overall scheme is called a predictor-corrector method iterated to convergence (see Sect. 2.6.3).

## 2.6.2  Chord Iteration

This iteration scheme will in general converge much more rapidly than the fixed-point iteration described above, if we use Newton's method to solve the nonlinear algebraic equations [33]. Rewriting equation (2.52) as:

$$g(y_{n+1}) = y_{n+1} - y_n - hf(x_{n+1}, y_{n+1}),  \tag{2.54}$$

we need to find the value of $y_{n+1}$ such that $g(y_{n+1}) \approx 0$.

Newton's method can be expressed, in a modified form, as:

$$y_{n+1}^{v+1} = y_{n+1}^v - \left(\frac{\partial g}{\partial y}\right)^{-1} g(y_{n+1}^v)$$

$$= y_{n+1}^v - \left(I - h\frac{\partial f}{\partial y}\right)^{-1} g(y_{n+1}^v)  \tag{2.55}$$

$$v = 0, 1, \ldots,$$

where the quantity $\partial f/\partial y$ is the *Jacobian matrix* evaluated at the point $y_{n+1}^v$, and $(I - h\partial f/\partial y)$ is called the *iteration matrix*. In practice this scheme is written as:

$$\left(I - h\frac{\partial f}{\partial y}\right)(y_{n+1}^{v+1} - y_{n+1}^v) = -g(y_{n+1}^v).  \tag{2.56}$$

This procedure will normally converge rapidly without a severe restriction on $h$. However it requires considerably more work for one iteration, as the Jacobian matrix should be generated and the iteration matrix decomposed to solve the equations.

Because of the computational expense, the Jacobian is not usually computed for each iteration or even at every time step, but only when certain tests indicate that the cost of computing a new Jacobian is justified by the improved performance in the nonlinear iteration.

Also, much efficiency can be gained by taking into account any sparsity structure of the Jacobian. This is because, not only can sparse Jacobian matrices be more readily created, but also efficient methods exist to decompose them. Indeed in practice we often do encounter sparse Jacobians and it is important to take advantage of this (e.g. Sect. 8.4).

## 2.6.3   Predictor-Corrector Methods

Using a predictor-corrector method is a convenient way to implement the implicit Adams-Moulton methods, where an initial guess of the new value is first "predicted" using the explicit method and then "corrected" using the implicit method.

To explain this we consider a two-step formula. Assuming that $y_n$ and $y_{n-1}$ are known, a predicted solution is computed first, using the explicit formula:

$$y_{n+1}^P = y_n + h[3/2 f_n - 1/2 f_{n-1}].$$
(2.57)

Then the derivative is evaluated at this point:

$$f_{n+1}^P = f(x_{n+1}, y_{n+1}^P),$$
(2.58)

and a corrected solution $y_{n+1}^C$ is computed at $x_{n+1}$, using the "implicit" method:

$$y_{n+1}^C = y_n + h/2[f(x_{n+1}, y_{n+1}^P) + f(x_n, y_n)].$$
(2.59)

This is usually referred to as *PEC* mode (Prediction, Evaluation, Correction). The values carried forward to the next iteration are $y_{n+1}^C, f_{n+1}^P$.

It is often more efficient after (2.59) has been computed to evaluate $f(x_{n+1}, y_{n+1}^C)$. The more accurate values $y_{n+1}^C, f_{n+1}^C$ are now carried forward and this is known as *PECE* mode.

As is clear from the above, the iteration is arranged in such a way that we do not need to solve implicit equations. However, this makes the formula explicit and, as mentioned earlier, this is a very poor choice for solving stiff problems but often an excellent choice for solving non-stiff problems.

## 2.6.4   Newton Iteration for Implicit Runge-Kutta Methods

The problem in implementing higher order implicit Runge-Kutta methods is that they are *heavily implicit* in the unknown $y$. If $y \in \mathfrak{R}^m$, an $s$-stage implicit Runge-Kutta method (2.23) requires a system of $sm$ nonlinear algebraic equations to be solved. The most efficient way of solving these equations is in general by a modified Newton iteration. If however we use a Gauss Runge-Kutta method then to solve for $y$ this requires $s^3 m^3/3$ multiplications per iteration. This compares badly for large $s$ with the $m^3/3$ multiplications required per iteration by multistep methods such as BDF and MEBDF. In fact this linear algebra cost is so high that Gauss methods are not used very extensively for solving stiff problems.

A class of methods which can be implemented efficiently if special care is taken with the linear algebra component of Newton's method are Radau methods as identified by Hairer and Wanner [20]. In particular the linear algebra cost for

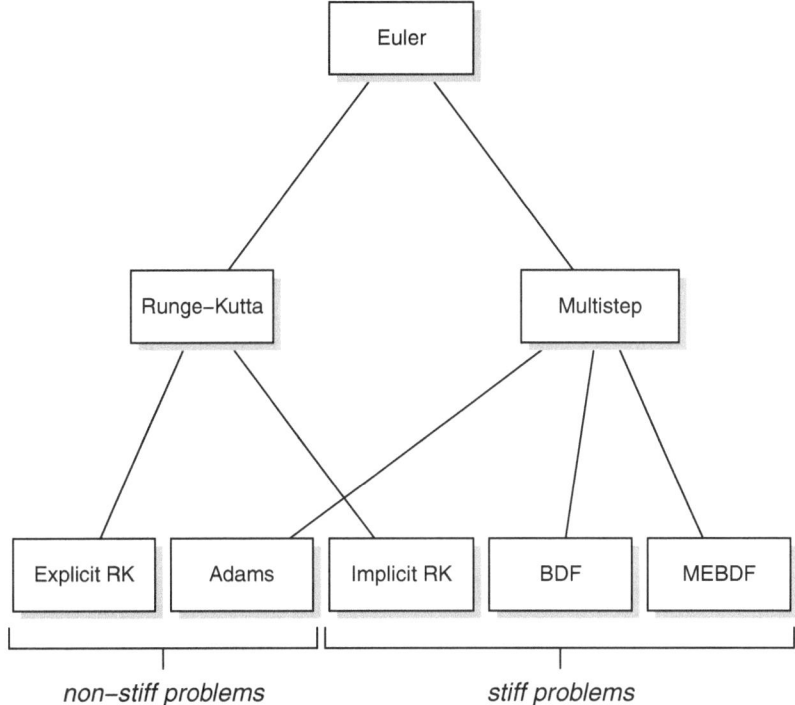

**Fig. 2.4**   Schematic representation of the main families of IVP solvers

Radau IIA methods is only $5m^3/3$ operations per iteration [20, p. 122]. However, a difficulty is that complex arithmetic is required. These methods, especially the fifth order Radau IIA method are very effective for solving stiff problems. Excellent accounts of implementation are given in [3, p. 48] and [20, p. 118].

## 2.7   Codes to Solve Initial Value Problems

The relationship between the various techniques for solving Initial Value Problems of ODEs is represented in Fig. 2.4. Explicit Runge-Kutta methods and Adams methods have comparable stability properties (see Figs. 2.1, 2.3), neither of them being well suited to solving stiff problems. For explicit Runge-Kutta methods, stability improves with order, but it is the other way round for Adams methods. Implicit Runge-Kutta methods, BDF, MEBDF and block-BVM are well suited to solving stiff problems mainly because of their excellent stability properties.

A list of solvers that are implemented in R can be found in the appendix.

## 2.7.1  Codes to Solve Non-stiff Problems

There are two main classes of formulae that are used for the numerical integration of non-stiff problems of ODEs, namely Adams methods and explicit Runge-Kutta methods. Adams codes have evolved from the well-known codes DVDQ of [26] and DIFSUB of [19]. Perhaps the two most widely used Adams methods at present are DE of Shampine and Gordon [39] which implements Adams methods in PECE mode and LSODE [23]. All these codes automatically adapt their step size, give a continuous solution and change the order of the method dynamically.

As regards explicit Runge-Kutta methods, some excellent codes are available. These codes provide an automatic step control procedure and give a continuous solution. However in general they do not attempt to change the order dynamically.

Among the most widely implemented explicit Runge-Kutta methods are the 3(2) pair of Bogacki and Shampine [6], the DOPRI5(4) due to Dormand and Prince [15, 16] and the Cash-Karp 5(4) formulae [12].

## 2.7.2  Codes to Solve Stiff Problems

Codes based on implicit Runge-Kutta methods, backward differentiation formulae, modified extended backward differentiation formulae and boundary value methods have proved to be very effective for the solution of stiff problems.

A widely used implicit Runge-Kutta code is RADAU5 [20]. BDF are the oldest and most widely used class of methods for solving stiff differential equations. Many people have made contributions to the efficient implementation of this class of formulae and perhaps the most well known codes are LSODE [23] and DASSL [30]. The stability properties of MEBDF formulae at high order make them generally the best candidates to solve very stiff problems and also to deal with problems when high accuracy is required. A well-known code implementing this is given in [11]. One code based on block-GAMs is GAM [24] and one based on A-stable block implicit methods of high order is BiMD [7] (see [41] for a description of block implicit methods).

## 2.7.3  Codes that Switch Between Stiff and Non-stiff Solvers

The Adams-BDF code LSODA [31] automatically switches from stiff to non-stiff and vice versa depending on the local stiffness properties of the problem. The novel feature of this code is that after switching from, say, the stiff to the non-stiff option it carries on with the integration. This is in contrast to the other algorithms which warn the user of stiffness and then stop.

Thus, when a problem is diagnosed as being stiff over the whole integration interval then LSODA will select a BDF formula of order 1–5 while it will use an

Adams method with order in the range 1–12 when a problem is detected as being non-stiff. For other problems the code will use both types of methods as the problem changes from being stiff to being non-stiff and back again. A code which uses a stiff method for stiff problems and a non-stiff method for non-stiff problems is called *type insensitive*. How to implement this approach is described in detail in [31]. Note in particular that LSODA uses norm bounds to estimate the dominant eigenvalue of the Jacobian (Sect. 2.5.2).

# References

1. Aceto, L., & Magherini, C. (2009). On the relations between $B_2$VMs and Runge-Kutta collocation methods. *Journal of Computational and Applied Mathematics, 231*(1), 11–23.
2. Ascher, U. M., & Petzold, L. R. (1998). *Computer methods for ordinary differential equations and differential-algebraic equations*. Philadelphia: SIAM.
3. Ascher, U. M., Mattheij, R. M. M., & Russell, R. D. (1995). *Numerical solution of boundary value problems for ordinary differential equations*. Philadelphia: SIAM.
4. Axelsson, A. O. H., & Verwer, J. G. (1985). Boundary value techniques for initial value problems in ordinary differential equations. *Mathematics of Computation, 45*(171), 153–171, S1–S4.
5. Björck, Å., & Dahlquist, G. (1974). *Numerical methods*. Englewood Cliffs: Prentice-Hall (Translated from the Swedish by Ned Anderson, Prentice-Hall Series in Automatic Computation).
6. Bogacki, P., & Shampine, L. F. (1989). A 3(2) pair of Runge–Kutta formulas. *Applied Mathematics Letters, 2*, 1–9.
7. Brugnano, L., & Magherini, C. (2004). The BiM code for the numerical solution of ODEs. *Journal of Computational and Applied Mathematics, 164–165*, 145–158.
8. Brugnano, L., & Trigiante, D. (1998). *Solving differential problems by multistep initial and boundary value methods: Vol. 6. Stability and control: Theory, methods and applications*. Amsterdam: Gordon and Breach Science Publishers.
9. Butcher, J. C. (1987). *The numerical analysis of ordinary differential equations, Runge–Kutta and general linear methods* (Vol. 2). Chichester/New York: Wiley.
10. Cash, J. R. (1980). On the integration of stiff systems of ODEs using extended backward differentiation formulae. *Numerische Mathematik, 34*, 235–246.
11. Cash, J. R., & Considine, S. (1992). An **MEBDF** code for stiff initial value problems. *ACM Transactions on Mathematical Software, 18*(2), 142–158.
12. Cash, J. R., & Karp, A. H. (1990). A variable order Runge–Kutta method for initial value problems with rapidly varying right-hand sides. *ACM Transactions on Mathematical Software, 16*, 201–222.
13. Curtiss, C. F., & Hirschfelder, J. O. (1952). Integration of stiff systems. *Proceedings of the National Academy of Science, 38*, 235–243.
14. Dahlquist, G. (1963). A special stability problem for linear multistep methods. *BIT, 3*, 27–43.
15. Dormand, J. R., & Prince, P. J. (1980). A family of embedded Runge–Kutta formulae. *Journal of Computational and Applied Mathematics, 6*, 19–26.
16. Dormand, J. R., & Prince, P. J. (1981). High order embedded Runge–Kutta formulae. *Journal of Computational and Applied Mathematics, 7*, 67–75.
17. Fehlberg, E. (1967). Klassische Runge–Kutta formeln funfter and siebenter ordnung mit schrittweiten-kontrolle. *Computing (Arch. Elektron. Rechnen), 4*, 93–106.
18. Fox, L. (1954). A note on the numerical integration of first-order differential equations. *The Quarterly Journal of Mechanics and Applied Mathematics, 7*, 367–378.

19. Gear, C. W. (1971). *Numerical initial value problems in ordinary differential equations.* Englewood Cliffs: Prentice-Hall.
20. Hairer, E., & Wanner, G. (1996). *Solving ordinary differential equations II: Stiff and differential-algebraic problems.* Heidelberg: Springer.
21. Hairer, E., Norsett, S. P., & Wanner, G. (2009). *Solving ordinary differential equations I: Nonstiff problems* (2nd rev. ed.). Heidelberg: Springer.
22. Henrici, P. (1962). *Discrete variable methods in ordinary differential equations.* New York: Wiley.
23. Hindmarsh, A. C. (1980). **LSODE** and **LSODI**, two new initial value ordinary differential equation solvers. *ACM-SIGNUM Newsletter, 15,* 10–11.
24. Iavernaro, F., & Mazzia, F. (1998). Solving ordinary differential equations by generalized Adams methods: Properties and implementation techniques. *Applied Numerical Mathematics, 28*(2–4), 107–126. (Eighth Conference on the Numerical Treatment of Differential Equations, Alexisbad, 1997).
25. Iavernaro, F., & Mazzia, F. (1999) Block-boundary value methods for the solution of ordinary differential equations. *SIAM Journal on Scientific Computing, 21*(1), 323–339 (electronic).
26. Krogh, F. T. (1969) **VODQ/SVDQ/DVDQ**, *Variable order integrators for the numerical solution of ordinary differential equations.* Pasadena, Calif: Jet Propulsion Laboratory.
27. Krogh, F. T. (1969). A variable step, variable order multistep method for the numerical solution of ordinary differential equations. In *Information processing 68 (Proceedings of the IFIP congress, Edinburgh, 1968): Vol. 1. Mathematics, software* (pp. 194–199). Amsterdam: North-Holland.
28. Kutta, W. (1901). Beitrag zur naeherungsweisen integration totaler differentialgleichungen. *Zeitschrift fur Mathematik und Physik, 46,* 435–453.
29. Lambert, J. D. (1973). *Computational methods in ordinary differential equations.* London/New York: Wiley.
30. Petzold, L. R. (1982). A description of **DASSL**: a differential/algebraic system solver. *IMACS Transactions on Scientific Computation.*
31. Petzold, L. R. (1983). Automatic selection of methods for solving stiff and nonstiff systems of ordinary differential equations. *SIAM Journal on Scientific and Statistical Computing, 4,* 136–148.
32. Press, W. H., Teukolsky, S. A., Vetterling, W. T., & Flannery, B. P. (1992) *Numerical recipes in FORTRAN. The art of scientific computing* (2nd ed.). New York: Cambridge University Press.
33. Press, W. H., Teukolsky, S. A., Vetterling, W. T., & Flannery, B. P. (2007) *Numerical recipes* (3rd ed.). Cambridge/New York: Cambridge University Press.
34. Runge, C. (1895). Ueber die numerische aufloesung von differentialgleichungen. *Mathematische Annalen, 46,* 167–178.
35. Rutishauser, H. (1952). uber die instabilitat von methoden zur integration gewohnlicher differentialgleichungen. *Zeitschrift fur angewandte Mathematik und Physik, 3,* 65–74.
36. Shampine, L. F. (1977). Stiffness and nonstiff differential equation solvers. ii detecting stiffness with Runge–Kutta methods. *ACM Transactions on Mathematical Software, 3,* 44–53.
37. Shampine, L. F. (1979). *Type-insensitive ODE codes based on implicit A-stable formulas* (pp. 79–244). Livermore: SAND, Sandia National Laboratories.
38. Shampine, L. F. (1980). Lipschitz constants and robust ODE codes. In *Computational methods in nonlinear mechanics* (pp. 47–449). Amsterdam: North Holland.
39. Shampine, L. F., & Gordon, M. K. (1975). *Computer solution of ordinary differential equations. The initial value problem.* San Francisco: W.H. Freeman.
40. Shampine, L. F., & Hiebert, K. L. (1977). Detecting stiffness with the Fehlberg (4, 5) formulas. *Computers and Mathematics with Applications, 3*(1), 41–46.
41. Shampine, L. F., & Watts, H. A. (1969). Block implicit one-step methods. *Mathematics of Computation, 23,* 731–740.
42. Skelboe, S. (1977). The control of order and steplength for backward differentiation methods. *BIT, 17,* 91–107.

# Chapter 3
# Solving Ordinary Differential Equations in R

**Abstract** Both Runge-Kutta and linear multistep methods are available to solve initial value problems for ordinary differential equations in the R packages **deSolve** and **deTestSet**. Nearly all of these solvers use adaptive step size control, some also control the order of the formula adaptively, or switch between different types of methods, depending on the local properties of the equations to be solved. We show how to trigger the various methods using a variety of applications pointing, where necessary, to problems that may arise. For instance, many practical applications involve discontinuities. As the integration routines assume that a solution is sufficiently differentiable over a time step, handing such discontinuities requires special consideration. We give examples of how we can implement a nonsmooth forcing term, switching behavior, and problems that include sudden jumps in the dependent variables. Since much computational efficiency can be gained by using the correct method for a particular problem, we end this chapter by providing a few guidelines as to how the most efficient solution method for a particular problem can be found.

## 3.1 Implementing Initial Value Problems in R

The R package **deSolve** [26] has several built-in functions for computing a numerical solution of initial value problems for ODEs.

They comprise methods to solve stiff and non-stiff problems, that deal with full, banded or arbitrarily sparse Jacobians etc... The methods included and the original source are listed in Sect. A.3.

A simplified form of the syntax for solving ODEs is:

```
ode(y, times, func, parms, ...)
```

where `times` holds the times at which output is wanted, `y` holds the initial conditions, `func` is the name of the R function that describes the differential equations, and `parms` contains the parameter values (or is `NULL`). Many additional inputs can be provided, e.g. the absolute and relative error tolerances (defaults

K. Soetaert et al., *Solving Differential Equations in R*, Use R!,
DOI 10.1007/978-3-642-28070-2_3, © Springer-Verlag Berlin Heidelberg 2012

rtol = 1e-6, atol = 1e-6), the maximal number of steps (maxsteps), the integration method etc. The default integration method is lsoda. If we type ?lsoda a help page is opened that contains a list of all options that can be changed. As all these options have a default value, we are not obliged to assign a value to them, as long as we are content with the default.

### 3.1.1  A Differential Equation Comprising One Variable

Ordinary differential equations are often used in population biology. One of the simplest equations describing population growth is the logistic equation [28]. This equation models the density changes of one species ($y$) in an environment where competition for available resources reduces population growth at high densities, and eventually leads to negative growth above a specific carrying capacity $K$. At very low density, the absence of competition allows exponential growth, at a growth rate $r > 0$:

$$y' = ry\left(1 - \frac{y}{K}\right),$$
$$y(0) = 2. \tag{3.1}$$

To implement this IVP in R we first define the two parameters, r and K, and the initial condition (yini) and assign values to them. The semi-colon ";" separates two statements; the "<-" is the assignment operator.

```
r <- 1; K <- 10; yini <- 2
```

The simple differential equation is implemented in an R function called derivs that takes as arguments the current time (t), the value of the dependent variable (y) and a parameter vector (parms), and that returns the derivative, as a list. The parameters $r$ and $K$, although defined outside of function derivs, are also known within the derivative function.[1]

```
derivs <- function(t, y, parms)
      list(r * y * (1-y/K))
```

We require output at 0.2 daily intervals for 20 days, which we specify in a vector (times); the R function seq creates the output time sequence. The model is solved, using R function ode. The integrator ode is available from the package **deSolve**, which is loaded first (library).

---

[1] A more robust but slightly more complex method is to put parameters in a parameter *vector*, and pass that to the derivative function via ode's argument parms. For models that are to be used to fit parameters to data, this is the most convenient way of passing parameters.

**Fig. 3.1** A simple initial value problem, solved twice with different initial conditions. See text for the R code

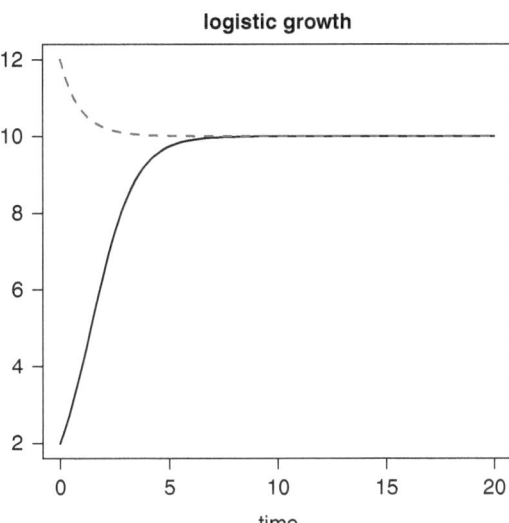

```
library(deSolve)
times <- seq(from = 0, to = 20, by = 0.2)
out    <- ode(y = yini, times = times, func = derivs,
                parms = NULL)
```

The model output in `out` is a `matrix` consisting of two columns, first `time`, then the state variable value `y`. We print the first three lines of this matrix:

```
head(out, n = 3)
```

```
       time        1
[1,]    0.0 2.000000
[2,]    0.2 2.339222
[3,]    0.4 2.716436
```

We now solve the differential equation with a different initial condition, $y(0) = 12$, and store the output in matrix `out2`:

```
yini <- 12
out2 <- ode(y = yini, times = times, func = derivs,
                parms = NULL)
```

The output of these two solutions is easily plotted, using the `deSolve`'s function `plot` with the solid lines twice as thick as the default (`lwd=2`) (Fig. 3.1). It shows for the first solution an initial fast increase of the density, levelling off towards the carrying capacity $K$. The second solution shows a gradual decrease towards $K$.

```
plot(out, out2, main = "logistic growth", lwd = 2)
```

## 3.1.2 Multiple Variables: The Lorenz Model

It is only slightly more complex to write a model that describes the dynamics of multiple variables.

The Lorenz equations [18] were the first chaotic dynamical system of ordinary differential equations to be described. They consist of three ordinary differential equations, expressing the dynamics of the variables, $X$, $Y$ and $Z$ that were assumed to represent idealized behavior of the Earth's atmosphere. The model equations are:

$$X' = aX + YZ,$$
$$Y' = b(Y - Z),$$
$$Z' = -XY + cY - Z, \tag{3.2}$$

where $X$, $Y$ and $Z$ refer to the horizontal and vertical temperature distribution and convective flow, and $a, b, c$ are parameters with values $-8/3$, $-10$ and $28$ respectively.

The R implementation starts by defining the parameter values and the initial condition. For the latter, we create a three-valued vector using the function "c ()", naming the elements "X", "Y" and "Z". These names are handy, as we can use them in the derivative function so making the code more readable, but they also serve to label the output (see below).

```
a    <- -8/3; b <- -10; c <- 28
yini <- c(X = 1, Y = 1, Z = 1)
```

Within the derivative function (called Lorenz), we make the names of the variables available by the construct with (as.list(y), {...}). Note that this statement effectively embraces all other statements within the function (i.e. the closing brackets "})" are the last before the curly braces terminating the function).

Similarly as in the previous example, the derivative function returns the derivatives, packed as a list, but now they are concatenated (c()) in a vector. Here it is *extremely* important [2] to return the values of the three derivatives, *in the same order* as in which the initial conditions are defined. With yini containing the values of variables X, Y and Z, the derivative vector should be returned as dX, dY, dZ.

```
Lorenz <- function (t, y, parms) {
  with(as.list(y), {
    dX <- a * X + Y * Z
    dY <- b * (Y - Z)
    dZ <- -X * Y + c * Y - Z
    list(c(dX, dY, dZ))    })
}
```

---

[2]it is the most common mistake that beginners make.

We solve the IVP for 100 days, producing output every 0.01 days; this small output step is necessary to obtain smooth figures. In general this does not affect the time step of integration; this is usually determined by the solver.

```
times <- seq(from = 0, to = 100, by = 0.01)
out   <- ode(y = yini, times = times, func = Lorenz,
             parms = NULL)
```

The output generated by the solvers from deSolve can conveniently be plotted using **deSolve**'s method plot. This works slightly differently from R's default plot method, as it depicts all variables at the same time, neatly arranged in several rows and columns. Using this plot method saves a lot of R statements, especially if the model contains many variables.

The first statement in the code section below plots the three dependent variables X, Y, Z in two rows and two columns. As we gave names to the initial conditions, the figures are correctly labeled (Fig. 3.2).

It is very simple to overrule this **deSolve**-specific plot function. By selecting specific variables from out (here "X" and "Y") rather than the entire output matrix, the default plotting method from R will be used. So, in the last statement, we depict variable Y versus X to generate the famous "butterfly" (Fig. 3.2).

```
plot(out, lwd = 2)
plot(out[,"X"], out[,"Y"], type = "l", xlab = "X",
   ylab = "Y", main = "butterfly")
```

## 3.2   Runge-Kutta Methods

The R package **deSolve** contains a large number of Runge-Kutta methods (Sect. 2.1). With the following statement all implemented methods are shown:

```
rkMethod()
```

```
 [1] "euler"    "rk2"      "rk4"      "rk23"     "rk23bs"
 [6] "rk34f"    "rk45f"    "rk45ck"   "rk45e"    "rk45dp6"
[11] "rk45dp7"  "rk78dp"   "rk78f"    "irk3r"    "irk5r"
[16] "irk4hh"   "irk6kb"   "irk4l"    "irk6l"    "ode23"
[21] "ode45"
```

They comprise simple Runge-Kutta formulae (Heun's method rk2, the classical fourth order Runge-Kutta, rk4) and several explicit Runge-Kutta pairs of orders 3(2) to orders 8(7). The embedded, explicit methods are according to Fehlberg [10] (rk..f), Dormand and Prince [8, 9] (rk..dp.), Bogacki and Shampine [3] (rk23bs, ode..) and Cash and Karp [7] (rk45ck).

The implicit Runge-Kutta's (irk..) from this list are not optimally coded; a better implicit Runge-Kutta is the (radau) method [11] that will be discussed in Chaps. 4 and 5.

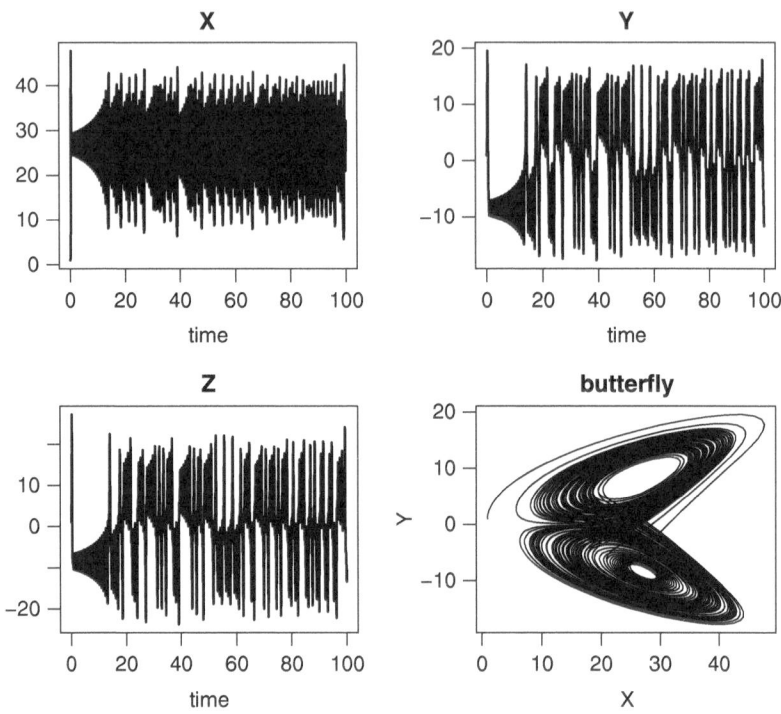

**Fig. 3.2** Solution of the Lorenz equation, a three-variable initial value problem that generates chaotic solutions. See text for the R code

Only two formulae (rk45dp7, rk45ck) support continuous solutions (Sect. 2.1.2.5).

The properties of a Runge-Kutta method can be displayed as follows:

```
rkMethod("rk23")
```

```
$ID
[1] "rk23"

$varstep
[1] TRUE

$FSAL
[1] FALSE

$A
     [,1] [,2] [,3]
[1,]  0.0    0    0
[2,]  0.5    0    0
[3,] -1.0    2    0

$b1
```

```
[1] 0 1 0

$b2
[1] 0.1666667 0.6666667 0.1666667

$c
[1] 0.0 0.5 2.0

$stage
[1] 3

$Qerr
[1] 2

attr(,"class")
[1] "list"      "rkMethod"
```

The output informs us whether the method uses a variable time step (`varstep`), and the first same as last (`FSAL`) strategy (this stores the derivative at the end of a step for use as the first evaluation in the next step). It gives the coefficients of the Butcher table (`A`, `b1`, `b2`, `c`, see Sect. 2.1.1, (2.5)), the number of function evaluations needed for one step (`stage`), and the order of the local truncation error (`Qerr`) of the method. If the method uses a variable step size the error is kept below a user defined relative and absolute tolerance, the default values for all the codes are `atol = 1e-6`, `rtol = 1e-6`. We note that, for some implementations, the vector `times` at which the output is wanted defines the mesh at which the method performs its steps, so the accuracy of the solution strongly depends on the input vector `times`.

It is also possible to define and use a new Runge-Kutta method (see help-file `?rkMethod`). Finally, in the R package **deTestSet** are some more Runge-Kutta methods, based on well-established FORTRAN codes (see Table A.7).

### 3.2.1 Rigid Body Equations

We first show how to use the Runge-Kutta methods by means of a standard test problem for non-stiff solvers, as proposed by Krogh, in [12]. It describes the Euler equations of a rigid body without external forces. The three dependent variables ($y_1$, $y_2$, $y_3$) are the coordinates of the rotation vector, while $I_1$, $I_2$, $I_3$ are the principal moments of inertia. The ODEs are:

$$
\begin{aligned}
y_1' &= \frac{(I_2 - I_3)}{I_1} y_2 y_3 \\
y_2' &= \frac{(I_3 - I_1)}{I_2} y_1 y_3 \\
y_3' &= \frac{(I_1 - I_2)}{I_3} y_1 y_2.
\end{aligned}
\tag{3.3}
$$

We implement the model with parameters $I_1 = 0.5, I_2 = 2, I_3 = 3$ and initial conditions $y_1(0) = 1, y_2(0) = 0, y_3(0) = 0.9$.

After loading the package **deSolve** and defining the initial conditions (yini), the model function is defined (rigidode). Although in previous examples, we made the code more readable by using the *names* of the variables, here we use their position in the variable vector y, instead. Also, the parameter values are hard-coded.

```
library(deSolve)
yini  <- c(1, 0, 0.9)
```

```
rigidode <- function(t, y, parms) {
    dy1 <- -2   * y[2] * y[3]
    dy2 <- 1.25* y[1] * y[3]
    dy3 <- -0.5* y[1] * y[2]
    list(c(dy1, dy2, dy3))
  }
```

The times at which output is wanted consists of a sequence of values, extending over 20 days, and at 0.01 day intervals. The ODEs are solved with the Cash-Karp Runge-Kutta method ("rk45ck"), and the first three rows of this matrix are visually inspected (head).

```
times <- seq(from = 0, to = 20, by = 0.01)
out   <- ode (times = times, y = yini, func = rigidode,
              parms = NULL, method = rkMethod("rk45ck"))
head (out, n = 3)
```

```
      time          1          2          3
[1,]  0.00 1.0000000 0.00000000 0.9000000
[2,]  0.01 0.9998988 0.01124950 0.8999719
[3,]  0.02 0.9995951 0.02249603 0.8998875
```

We could plot the three state variables, using **deSolve**'s plot method as in the previous example. However, it is more instructive to plot all variables in one figure instead (Fig. 3.3). We use R function matplot to do so. Rather than using the default settings of the function (points), we choose solid lines (type = "l", lty = "solid"), twice as thick as the default (lwd = 2) and with different colour for each state variable. The first column of out holds the time and is used as the x-variable, while all except the first column (out[,-1]) are used as y-variables.

```
matplot(x = out[,1], y = out[,-1], type = "l", lwd = 2,
        lty = "solid", col = c("red", "blue", "black"),
        xlab = "time", ylab = "y", main = "rigidode")
legend("bottomright", col = c("red", "blue", "black"),
        legend = c("y1", "y2", "y3"), lwd = 2)
```

Another way of depicting the output is to plot the three coordinates of the rotation vector in a 3D plot. This can easily be done using the R package **scatterplot3d** [17].

```
library(scatterplot3d)
scatterplot3d(out[,-1], type = "l", lwd = 2, xlab = "",
              ylab = "", zlab = "", main = "rigidode")
```

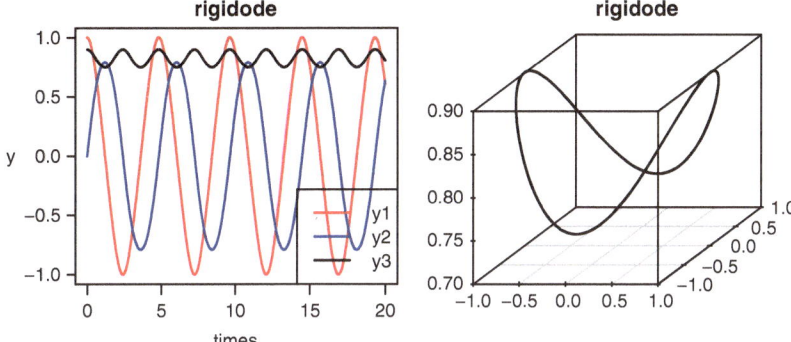

**Fig. 3.3** The Euler equations of a rigid body without external forces, solved with the Cash-Karp 5(4) Runge-Kutta formula. See text for the R code

### 3.2.2 Arenstorf Orbits

Our next example, the Arenstorf problem, is from Astronomy, and describes the movement of a small body orbiting regularly around two larger objects, such as a spacecraft going between the Earth and the Moon. The two large bodies have mass $m_1$ and $m_2$ and move in a circular rotation (coordinates $y_1$ and $y_2$) in a plane, while the third body has negligible mass and is moving in the same plane.

It was necessary to solve this problem in order to determine the path that the Apollo spacecraft had to take in its journey between the Earth and the Moon. The problem was solved by Arenstorf [1] and now it is an often used test problem for non-stiff solvers.

If we define $\mu_1 = \dfrac{m_1}{m_1 + m_2}$ and $\mu_2 = 1 - \mu_1$, then the equations are [1]:

$$\begin{aligned}
y_1'' &= y_1 + 2y_2' - \mu_2 \frac{y_1 + \mu_1}{D_1} - \mu_1 \frac{y_1 - \mu_2}{D_2} \\
y_2'' &= y_2 - 2y_1' - \mu_2 \frac{y_2}{D_1} - \mu_1 \frac{y_2}{D_2} \\
D_1 &= ((y_1 + \mu_1)^2 + y_2^2)^{(3/2)} \\
D_2 &= ((y_1 - \mu_2)^2 + y_2^2)^{(3/2)},
\end{aligned}$$
(3.4)

where $\mu_1 = 0.012277471$. For certain values of the initial conditions, this problem has a periodic solution. One set of initial conditions with this property is: $y_1(0) = 0.994, y_2(0) = 0, y_1'(0) = 0, y_2'(0) = -2.00158510637908252240537862224$.

Before solving these equations, we expand the second order equations in two first order ones ($y_3 = y_1'$ and $y_4 = y_2'$)

```
library(deSolve)
Arenstorf <- function(t, y, p) {
    D1 <- ((y[1]  + mu1)^2 + y[2]^2)^(3/2)
    D2 <- ((y[1]  - mu2)^2 + y[2]^2)^(3/2)
    dy1 <- y[3]
    dy2 <- y[4]
    dy3 <- y[1] + 2*y[4] - mu2*(y[1]+mu1)/D1 - mu1*(y[1]-mu2)/D2
    dy4 <- y[2] - 2*y[3] - mu2*y[2]/D1 - mu1*y[2]/D2
    return(list( c(dy1, dy2, dy3, dy4) ))
  }
mu1    <- 0.012277471
mu2    <- 1 - mu1
yini   <- c(y1 = 0.994, y2 = 0,
            dy1 = 0, dy2 = -2.00158510637908252240537862224)
times <- seq(from = 0, to = 18, by = 0.01)
```

We solve the above IVP with the fifth order Dormand and Prince method (DOPRI5(4) [8]):

```
out <- ode(func = Arenstorf, y = yini, times = times,
           parms = 0, method = "ode45")
```

We can also solve the same problem with a second and third set of initial conditions:

```
yini2 <- c(y1 = 0.994, y2 = 0,
           dy1 = 0, dy2 = -2.03173262955733368357302057924)
out2 <- ode(func = Arenstorf, y = yini2, times = times,
            parms = 0, method = "ode45")
yini3 <- c(y1 = 1.2, y2 = 0,
           dy1 = 0, dy2 = -1.049357510)
out3 <- ode(func = Arenstorf, y = yini3, times = times,
            parms = 0, method = "ode45")
```

We end by first plotting the first two variables versus time for all three solutions, arranging the figures in two rows and two columns (mfrow). Then we plot the trajectories of the three runs (Fig. 3.4).

```
plot(out, out2, out3, which = c("y1", "y2"),
     mfrow = c(2, 2), col = "black", lwd = 2)
plot(out[ ,c("y1", "y2")], type = "l", lwd = 2,
     xlab = "y1", ylab = "y2", main = "solutions 1,2")
lines(out2[ ,c("y1", "y2")], lwd = 2, lty = 2)
plot(out3[ ,c("y1", "y2")], type = "l", lwd = 2, lty = 3,
     xlab = "y1", ylab = "y2", main = "solution 3")
```

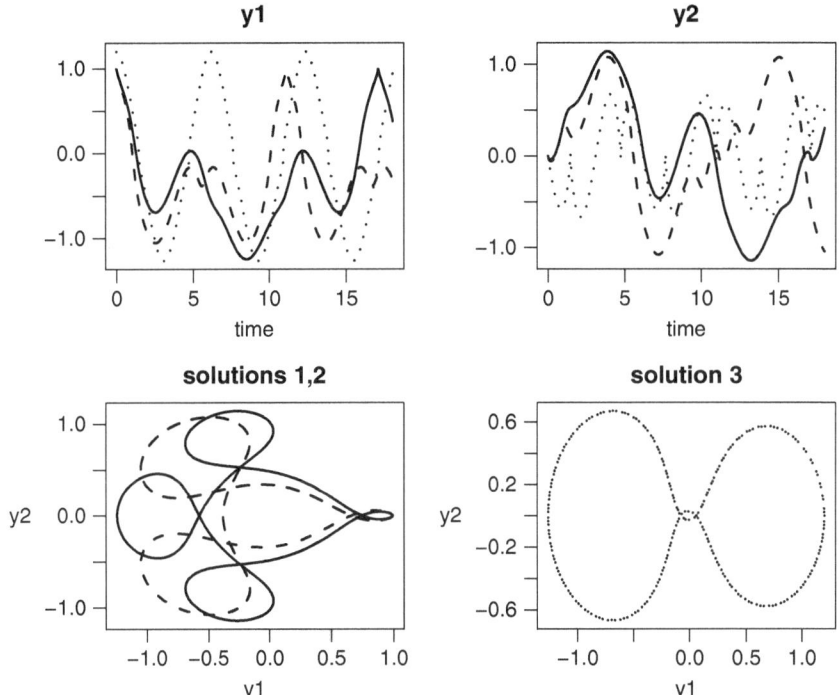

**Fig. 3.4** The Arenstorf problem, solved with the dopri5 Runge-Kutta method. See text for the R code

## 3.3  Linear Multistep Methods

The solvers vode [4], lsode [13], and lsodes [14] from the R package **deSolve** implement both variable-coefficient Adams methods as well as backward differentiation formulas (the default) of variable order. The Adams methods are better suited for non-stiff problems, the BDF for stiff problems. In these codes, the maximal order for the Adams and BDF methods are 12 and 5 respectively. Whereas in the above-mentioned solvers, it is left to the user to specify whether to use a stiff or non-stiff method, the solver lsoda [19] will detect whether stiffness is present or not, and trigger an appropriate change in the solution method if this property of stiffness changes (see Sect. 2.7.3). This is such a robust procedure that lsoda is the default integration method chosen by ode.

In some cases, one may find it more efficient to select another integration method rather than this default. Two implementations of the Adams methods are available, one that uses a predictor corrector implementation with a functional iteration as the corrector (called "adams"), and a second that implements the implicit Adams method ("impAdams") by solving the implicit equation using chord iteration based

on the Jacobian (see Sect. 2.6). It is simplest to use the function ode with the
appropriate method, to trigger a specific multistep method. For instance,

```
ode(y, times, func, parms, ...)
ode(y, times, func, parms, method = "bdf", ...)
ode(y, times, func, parms, method = "adams", ...)
ode(y, times, func, parms, method = "impAdams", ...)
ode(y, times, func, parms, method = vode, ...)
```

will use lsoda (the default method), the backward differentiation formula, and
the simple Adams and implicit Adams method (based on the code lsode), and the
method vode respectively. Note that it is allowed to pass both the name of the
function, or the function itself.

Finally, more multistep methods are available from the package **deTestSet**: functions gamd [16] (implementing the generalized Adams methods), mebdfi [6] (the
modified extended backward differentiation formula) and bimd[5] (implementing
block implicit methods). The calling sequence is:

```
library(deTestSet)
ode(y, times, func, parms, method = gamd, ...)
ode(y, times, func, parms, method = mebdfi, ...)
ode(y, times, func, parms, method = bimd, ...)
```

or:

```
gamd(y, times, func, parms, ...)
mebdfi(y, times, func, parms,...)
bimd(y, times, func, parms,...)
```

These codes will be extensively used when we deal with solving DAEs, in
Chap. 5.

### 3.3.1   Seven Moving Stars

The Pleiades problem [12] is a celestial mechanics problem of seven stars, with
masses $m_i$, in the two-dimensional plane of coordinates $(x, y)$. The stars are
considered to be point masses.

The only force acting on them is gravitational attraction, with gravitational
constant $G$ (units of $m^3 kg^{-1} s^{-2}$).

If $r_{ij} = (x_i - x_j)^2 + (y_i - y_j)^2$ is the square of the distance between stars $i$ and $j$,
then the second order equations describing their movement are given by:

$$
\begin{aligned}
x_i'' &= G \sum_{j \neq i} m_j \frac{(x_j - x_i)}{r_{ij}^{3/2}} \\
y_i'' &= G \sum_{j \neq i} m_j \frac{(y_j - y_i)}{r_{ij}^{3/2}},
\end{aligned}
\tag{3.5}
$$

where, to estimate the acceleration of star $i$, the sum is taken over all the interactions with the other stars $j$. Written as first order ODEs, we obtain:

$$x_i' = u_i$$
$$y_i' = v_i$$
$$u_i' = G \sum_{j \neq i} m_j \frac{(x_j - x_i)}{r_{ij}^{3/2}} \qquad (3.6)$$
$$v_i' = G \sum_{j \neq i} m_j \frac{(y_j - y_i)}{r_{ij}^{3/2}},$$

where $x_i, u_i, y_i, v_i$ are the positions and velocities in the $x$ and $y$ directions of star $i$ respectively.

With 7 stars, and 4 differential equations per star, this problem comprises 28 equations. As in [12], we assume that the masses $m_i = i$ and that the gravitational constant $G$ equals 1; the initial conditions are found in [12]. We integrate the problem in the time interval $[0, 3]$.

In the function that implements the derivative in R (pleiade), we start by separating the input vector Y into the coordinates (x, y) and velocities (u, v) of each star.

The distances in the $x$ and $y$ directions are created using R function outer. This function will apply FUN for each combination of x and y. It thus creates a matrix with seven rows and seven columns, having for distx on the position $i, j$, the value $x_i - x_j$.

The matrix containing the values $r_{ij}^{3/2}$, called rij3 is then calculated based on distx and disty.

Finally we multiply matrix distx or disty with the vector containing the masses of the stars (starMass), and divide by matrix rij3.

The result of these calculations are two matrices (fx, fy), with seven rows and columns. As the distance between a body and itself is equal to 0, this matrix has NaN (Not a Number) on the diagonal.

The required summation to obtain $u'$ and $v'$ (3.5) is done using R function colSums; the argument na.rm = TRUE ensures that these sums ignore the NaNs on the diagonal of fx and fy.

During the movement of the seven bodies several quasi-collisions occur, where the squared distance between two bodies are as small as $10^{-3}$. When that happens, the accelerations $u', v'$ get very high.

Thus, over the entire integration interval, there are periods with slow motion and periods of rapid motion, such that this problem can only be efficiently solved with an integrator that uses adaptive time stepping.

As the problem is non-stiff, it is solved with the "adams" method. We use the function system.time to have information about the elapsed time required to obtain the solution.

```
library(deSolve)
pleiade <- function (t, Y, pars) {
    x <- Y[1:7]
    y <- Y[8:14]
    u <- Y[15:21]
    v <- Y[22:28]

    distx <- outer(x, x, FUN = function(x, y) x - y)
    disty <- outer(y, y, FUN = function(x, y) x - y)

    rij3 <- (distx^2 + disty^2)^(3/2)

    fx <- starMass * distx / rij3
    fy <- starMass * disty / rij3

    list(c(dx = u,
           dy = v,
           du = colSums(fx, na.rm = TRUE),
           dv = colSums(fy, na.rm = TRUE)))
 }
starMass <- 1:7
yini<- c(x1= 3, x2= 3, x3=-1, x4=-3,    x5= 2, x6=-2,    x7= 2,
         y1= 3, y2=-3, y3= 2, y4= 0,    y5= 0, y6=-4,    y7= 4,
         u1= 0, u2= 0, u3= 0, u4= 0,    u5= 0, u6=1.75, u7=-1.5,
         v1= 0, v2= 0, v3= 0, v4=-1.25, v5= 1, v6= 0,    v7= 0)
print(system.time(
 out <- ode(func = pleiade, parms = NULL, y = yini,
            method = "adams", times = seq(0, 3, 0.01))))
```

```
   user   system elapsed
   0.12     0.00     0.13
```

To use the solvers in the **deTestSet** package the instructions are:

```
library(deTestSet)
print(system.time(
 out2 <- ode(func = pleiade, parms = NULL, y = yini,
             method = bimd, times = seq(0, 3, 0.01))))
```

```
   user   system elapsed
    1.3      0.0      1.3
```

```
print(system.time(
 out3 <- ode(func = pleiade, parms = NULL, y = yini,
             method = mebdfi, times = seq(0, 3, 0.01))))
```

```
   user   system elapsed
   0.86     0.00     0.89
```

The execution times of bimd, gamd, mebdfi show that the Adams method is the most efficient solver for this problem. To create Fig. 3.5, we first loop over the seven stars, plotting the trajectory of each star in a separate figure. The initial value

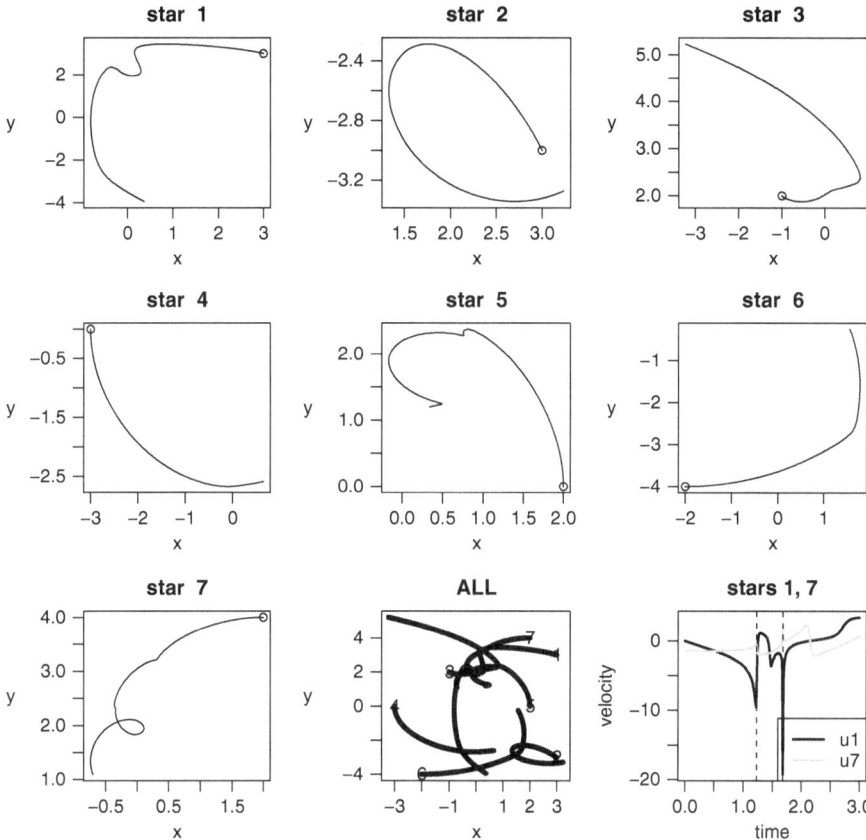

**Fig. 3.5** The pleiades problem describes seven stars moving in a 2-D plane. The first eight figures depict the trajectory of each star; in the last figure the velocity in the *x*-direction of *stars 1* and *7* are depicted versus time; there is a near-collision at $t = 1.23$ and $t = 1.68$, indicated with a *dotted vertical line*. See text for the R code

is represented as a `point`. Then we plot all trajectories in one figure, labelling the initial point of each star (`text`) with its number. The last figure depicts the velocity *u* of the first and seventh star versus time. These two stars almost collide at $t = 1.23$ and at $t = 1.68$. We emphasise this by drawing a vertical line at these times; `abline(v =...)` does this.

```
par(mfrow = c(3, 3), mar = c(4, 4, 3, 2))
for (i in 1:7) {
   plot(out[,i+1], out[,i+8], type = "l",
        main = paste("star ",i), xlab = "x", ylab = "y")
   points (yini[i], yini[i+7])
 }
#
plot(out[, 2:8], out[, 9:15], type = "p", cex = 0.5,
```

```
      main = "ALL", xlab = "x", ylab = "y")
text(yini[1:7], yini[8:14], 1:7)
#
matplot(out[,"time"], out[, c("u1", "u7")], type = "l",
    lwd = 2, col = c("black", "grey"), lty = 1,
    xlab = "time", ylab = "velocity", main = "stars 1, 7")
abline(v = c(1.23, 1.68), lty = 2)
legend("bottomright", col = c("black", "grey"), lwd = 2,
        legend = c("u1", "u7"))
```

### 3.3.2  A Stiff Chemical Example

We now implement an example, slightly adapted from [15], describing ozone
concentrations in the atmosphere. This example serves two purposes: (1) it provides
a stiff problem (see Sect. 2.5) and (2) we use it to demonstrate how to use external
data in a differential equation model.

The model describes the following three chemical reactions between oxygen
($O_2$), ozone ($O_3$), atomic oxygen ($O$), nitrogen oxide ($NO$), and nitrogen dioxide
($NO_2$):

$$
\begin{aligned}
NO_2 + hv \ &\xrightarrow{r_1(t)}\ NO + O \\
O + O_2 \ &\xrightarrow{r_2}\ O_3 \\
NO + O_3 \ &\xrightarrow{r_3}\ O_2 + NO_2.
\end{aligned}
\tag{3.7}
$$

The first reaction is the photo-dissociation of $NO_2$ to form $NO$ and $O$. This reaction
depends on solar radiation ($hv$), and therefore its rate ($r_1(t)$) changes drastically at
sunrise and sunset. The second reaction describes the production of ozone, which
proceeds at a rate $= r_2$. In the third reaction, $NO$ reacts with ozone (rate $r_3$).

The Earth's ozone levels are of great interest as at high concentrations it is
harmful to humans and animals, and because ozone is also a green-house gas.

According to the mass action law [2], the speed of the reaction is proportional to
the product of the concentrations of the participating molecules. Thus, for $r_1$, $r_2$ and
$r_3$ we can write:

$$
\begin{aligned}
r_1(t) &= k_1(t)[NO_2] \\
r_2 &= k_2[O] \\
r_3 &= k_3[NO][O_3].
\end{aligned}
\tag{3.8}
$$

For the derivation of $r_2$ we assumed the oxygen concentration to be constant, which,
in the Earth's atmosphere is not too crude an assumption.

Based on these rates, the differential equations expressing the dynamics for the
concentrations of $O$, $NO$, $NO_2$, and $O_3$ (here written as $([O], \ldots, [O_3])$ are [15]:

$$
\begin{aligned}
{[O]}' &= k_1(t)[NO_2] - k_2[O] \\
{[NO]}' &= k_1(t)[NO_2] - k_3[NO][O_3] + \sigma \\
{[NO_2]}' &= k_3[NO][O_3] - k_1(t)[NO_2] \\
{[O_3]}' &= k_2[O] - k_3[NO][O_3].
\end{aligned}
\tag{3.9}
$$

Here $\sigma$ is the emission rate of nitrogen oxide, which we assume constant, while the reaction rate $k_1$ depends linearly on the solar radiation, $hv(t)$, according to:

$$k_1(t) = k_{1a} + k_{1b}hv(t). \tag{3.10}$$

As the solar radiation is not constant, this rate changes with time. In the next section we will show how to efficiently implement the solar radiation into this model.

### 3.3.2.1 External Variables

Often external variables such as the solar radiation are imposed on a differential equation problem by means of a *time series*. For the ozone chemistry example, the solar radiation time series consists of one measurement taken each 0.5 h, and extending over 5 days. Such data is best input as a data.frame or a matrix. Here the data.frame called Light is read from a file ("Light.rda"); the first four datapoints are shown:

```
load(file = "Light.rda")
head(Light, n = 4)
```

```
        day     irrad
1 0.0000000   0.0000
2 0.3333333   0.0000
3 0.3541667 164.2443
4 0.3750000 204.7486
```

In order to use this data in the ODE system, we need a way to interpolate the half-hourly observations to the exact time points at which the integration routine will require them. However, as the solvers adapt their time steps depending on local properties of the integration, we have no prior knowledge about the times at which the derivative function will be called.

R function approxfun is an ingenious method that allows a user to perform this interpolation. It is used in two steps:

1. First an *interpolating function* that contains the data (*x*- and *y*-values) is constructed.
2. This function is then used to provide the interpolated value at intermediate time steps.

The interpolating function is created by a call to approxfun, passing the *x*- and *y*-data in data.frame Light. We want the values to be linearly interpolated in between data points, but as this is the default interpolation method of approxfun, we do not need to specify this.

```
irradiance <- approxfun(Light)
```

The interpolating function is called irradiance here, and it is created only once, outside of the derivative function (chemistry, see below). Once created, we can simply call function `irradiance` with the appropriate time-value to retrieve the solar radiation at that time. To show that this actually works, the next statement calculates the irradiances at specific time points $(0, 0.25, 0.5, 0.75, 1)$:

```
irradiance(seq(from = 0, to = 1, by = 0.25))
```

```
[1]    0.0000    0.0000 698.8911 490.4644    0.0000
```

Within the derivative function, we will use `irradiance` to interpolate the time series to the requested time of the simulation (as given by input argument `t`).

We are now well equipped to write the R code for the chemistry model; we define parameter values and initial conditions first.[3]

```
k3 <- 1e-11; k2 <- 1e10; k1a <- 1e-30
k1b <- 1; sigma <- 1e11
yini <- c(O = 0, NO = 1.3e8, NO2 = 5e11, O3 = 8e11)
```

The derivative function is defined next; it not only returns the derivatives, but also the solar radiation (last statement).

```
chemistry <- function(t, y, parms) {
   with(as.list(y), {

      radiation <- irradiance(t)
      k1   <- k1a + k1b*radiation

      dO    <-   k1*NO2  - k2*O
      dNO   <-   k1*NO2          - k3*NO*O3 + sigma
      dNO2  <- -k1*NO2           + k3*NO*O3
      dO3   <-           k2*O - k3*NO*O3
      list(c(dO, dNO, dNO2, dO3), radiation = radiation)
   })
}
```

Note how, in the first statement of function `chemistry`, the light intensity (or `radiation`) at time `t` is extracted by a call to the interpolating function `irradiance`.

We solve the IVP over a period of 5 days using the "`bdf`" method . For a model that is stiff this method is very efficient.

```
times <- seq(from = 0, to = 5, by = 0.01)
out <- ode(func = chemistry, parms = NULL, y = yini,
           times = times, method = "bdf")
```

We use **deSolve**'s `plot` method to plot all dependent variables and the output variable `radiation` in one figure.

---

[3]Here it is worthwhile to point to the difference of the letter "O" and the number "0" in the definition of `yini`; many strange behaviors of DE models are due to mistyping O and 0.

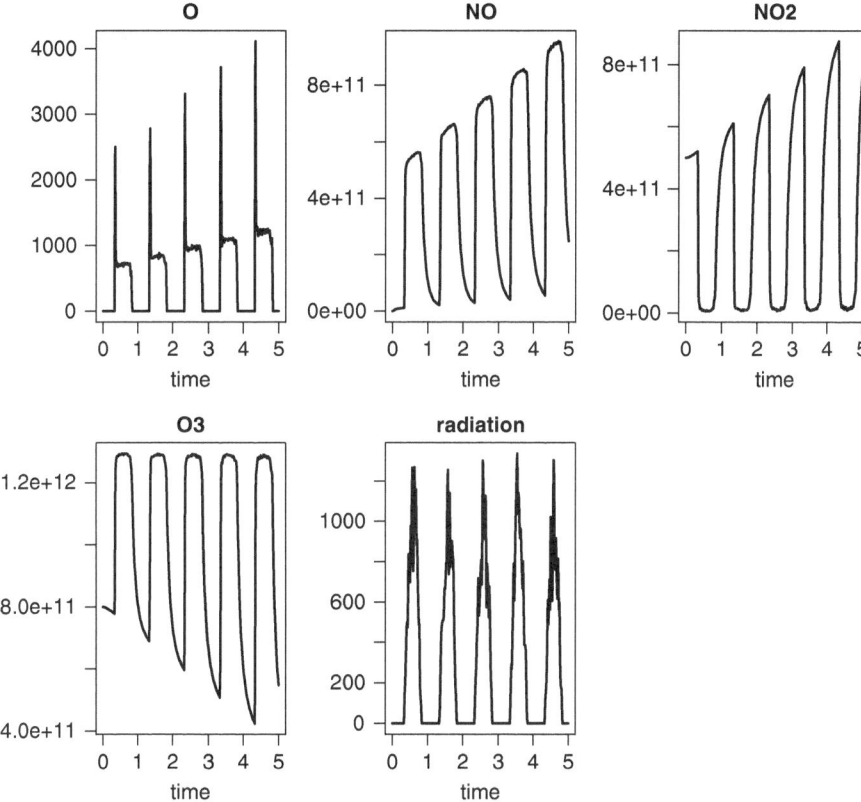

**Fig. 3.6** The atmospheric chemistry model solved with the bdf method. See text for the R code

```
plot(out, type = "l", lwd = 2 )
```

The results (Fig. 3.6) show how, at the onset of the day the $O$ and $NO$ concentrations increase drastically, due to the photo-dissociation reaction, which rapidly exhausts $NO_2$. As most of the $O$ produced reacts with $O_2$ at a very high rate to form $O_3$, the $O$ concentrations increase only little compared to $NO$.

## 3.4  Discontinuous Equations, Events

Many real-world model applications involve discontinuities. As R's integration routines all assume that a solution is sufficiently continuous over a time step, handling such discontinuities sometimes requires special consideration.

There are several levels of difficulty arising in discontinuous model systems. In the simplest case, it is just the forcing or external variables of the system that are not smooth. We gave an example of that in the previous section, where ozone degradation depended on light which was prescribed to the model by linear interpolation between data points.

In this section we give several other examples. The first is a (pharmacokinetic) example of a patient taking a pill every day. This changes the dosing of the drug in the blood in a discontinuous way. As these discontinuities affect the *derivatives* of the dependent variables they are quite easy to handle.

It is much more difficult to deal with events that cause sudden jumps in the *values* of the dependent variables. This is because the integration methods ignore all direct changes to the state variable values if they occur within the derivative function. We give an example of a patient injecting a drug at regular intervals.

In the above two examples, it is known in advance *when* the change will be triggered, as they occur at preset times. It is even more difficult to deal with sudden changes that occur only when certain conditions are met. In such cases, a *root* function is necessary to locate when these conditions arise, after which an *event* function is called to perform the change. We exemplify this type of discontinuity with an ODE describing a bouncing ball, and a model that describes temperature changes in a heat-controlled room.

Finally, it is not uncommon for solvers that take large steps to miss certain events. As this leads to wrong solutions, it is important to recognise this, and to take appropriate action to avoid it happening.

## 3.4.1  Pharmacokinetic Models

In order to be effective, the concentration of a drug taken by a patient must be large enough, yet too high concentrations may have serious side effects. Pharmacokinetic models are non-pervasive tools to test the optimal frequency and dosing of drug intake. They represent absorption, distribution, decay and excretion of a drug [21].

Drugs can be dosed orally (pills), or directly injected in the blood. In the first case, the action will operate on the processes (absorption through the gut),while in the latter case, the action will (almost) instantaneously alter the concentration in the blood.

### 3.4.1.1  A Two-Compartment Model Describing Oral Drug Intake

Consider a patient taking a pill every day at the same time. As the pill passes the gastro-intestinal tract, the drug enters the blood by absorption through the gut wall. The delivery of the drug to the gastro-intestinal tract proceeds for 1 h after which it ceases until the next ingestion and so on.

Once in the blood, the drug distributes in the tissues, where it is chemically inactivated, so that it can be excreted from the body. An (overly) simple two-compartment model, representing drug concentration in the gut ($y_1$) and in the blood ($y_2$) can represent this process [24]:

$$y_1' = -ay_1 + u(t)$$
$$y_2' = ay_1 - by_2. \tag{3.11}$$

Here $a$ is the absorption rate, $b$ is the removal rate from the blood, and the term $u(t)$ represents the daily delivery of the drug to the intestinal tract, which we assume to occur over a period of 1 h.

The discontinuity in this model lies in the dosing of the drug to the intestine ($u(t)$), which takes a constant value for 1 h, after which it is 0 for the rest of the day.

We now implement the R code for this pharmacokinetic problem. We first define parameters and initial conditions (starting with 0 concentration in both the intestinal tract and blood) and then implement the derivative function (pharmacokinetics).

As the uptake is periodic, we can use the modulo function (%%) to represent the uptake of the drug:

```
a <- 6; b <- 0.6
yini <- c(intestine = 0, blood = 0)
```

```
pharmacokinetics <- function(t, y, p) {
    if ( (24*t) %% 24 <= 1)
       uptake <- 2
    else
       uptake <- 0
    dy1 <- - a* y[1] + uptake
    dy2 <-   a* y[1] - b *y[2]
    list(c(dy1, dy2))
  }
```

The problem is solved in the usual way, and its output plotted:

```
times <- seq(from = 0, to = 10, by = 1/24)
out <- ode(func = pharmacokinetics, times = times,
           y = yini, parms = NULL)
plot(out, lwd = 2, xlab = "day")
```

The upper panel of Fig. 3.7 shows the result. At the start of the solution, and for each first hour of the day, the drug is ingested which causes a steep rise in the intestinal concentrations. As the drug enters the blood, its concentration in the intestine decreases exponentially, while initially increasing in the blood, where it is degraded. Since the inflow to the blood drops exponentially; at a certain point in time loss will exceed input and the concentration in the blood will start to decrease until the next drug dose.

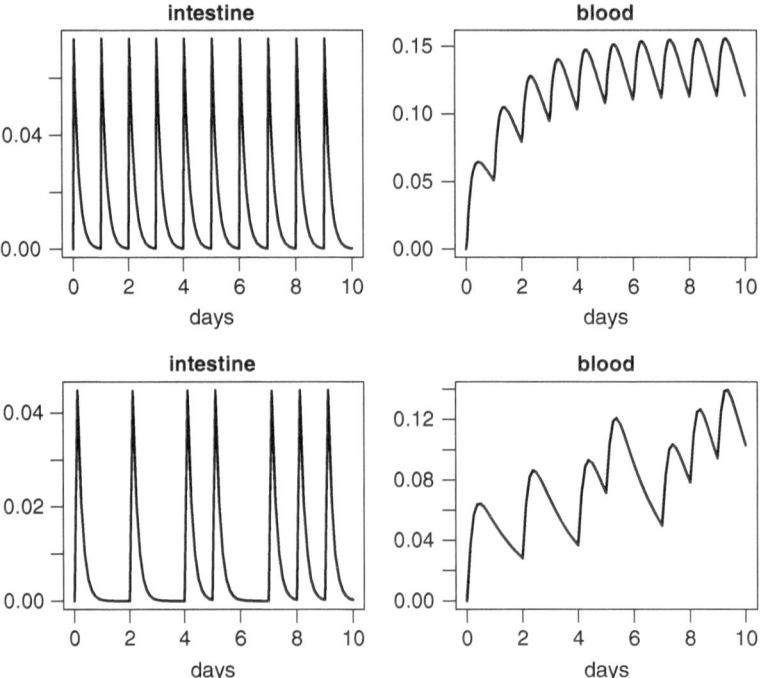

**Fig. 3.7** The 2-compartment pharmacokinetic model, describing the concentration of a daily-dosed drug in the intestinal tract and in the blood. *Above*: correct solution; *below*: wrong solution caused by too "efficient" a solver. See text for the R code

The initial concentration in the blood is very small but, as time proceeds, the daily-averaged concentration increases to reach some kind of dynamic equilibrium, which is nearly attained after 5–6 days.

When dealing with models that have such discontinuities it is very easy to miss some of the short inputs or abrupt changes. As it is important to recognise such a failure of the solvers, we will trigger one now. We ask for less output, say every 3 h, and we use an efficient integration routine that can take large time steps, the implicit Adams method (`"impAdams"`).

```
times <- seq(0, 10,  by = 3/24)
out2 <- ode(func = pharmacokinetics, times = times,
            y = yini, parms = NULL, method = "impAdams")
plot(out2, lwd = 2, xlab = "days")
```

The results (Fig. 3.7 lower panel) clearly show that the behavior is correct only during the first day of the simulation. After that, the integrator regularly misses the drug pulse! As the algorithm selects its time step according to the local accuracy requirements, before the intake of the drug the time step is rather large, occasionally much larger than the pulse width of 1 h, during which the drug is taken.

Consequently it may easily miss this pulse until, by chance, it steps into another pulse interval.

When this kind of behavior is suspected, it is wise to restrict the size of the time step. For this particular model, adding argument `hmax = 1/24` or using a lower absolute tolerance `atol = 1e-10` in the call to the `ode` function will fix the problem.

### 3.4.1.2  A One-Compartment Model Describing Drug Injection

In the previous example, uptake of a pill changed the *derivative* of the intestinal drug concentration. The differential equations differ when the drug is injected directly in the blood stream. In this case, the *concentration* of the drug in the blood is quasi-instantaneously altered, and there is no need to describe the concentration in the intestinal tract. The model that describes the dynamics of the drug in the blood, *in between injections* reads:

```
b    <- 0.6
yini <- c(blood = 0)
```

```
pharmaco2 <- function(t, blood, p) {
    dblood <-   - b * blood
    list(dblood)
  }
```

Assume a patient who injects daily doses of a drug in her veins, each time increasing the concentration by 40 units. The injection event causes the *value* of the state variable to be altered not the *derivative*, as in previous example. Unfortunately, the solvers in **deSolve** ignore any changes in the state variable values when made in the derivative function, so this is not so simple to implement.

The drug injections have to be specified in a special event `data.frame`

```
injectevents <- data.frame(var = "blood",
                           time =   0:20,
                           value = 40,
                           method = "add")
head(injectevents)
```

```
    var time value method
1 blood    0    40    add
2 blood    1    40    add
3 blood    2    40    add
4 blood    3    40    add
5 blood    4    40    add
6 blood    5    40    add
```

**Fig. 3.8** The 1-compartment
pharmacokinetic model,
describing the concentration
of a daily-dosed drug injected
directly in the blood stream.
See text for the R code

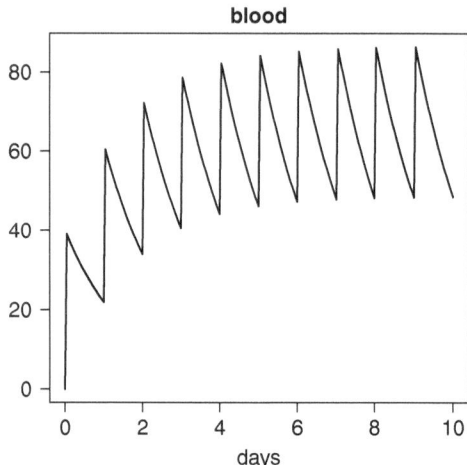

The event is said to add the *value* 40 to the *var*iable blood, at the prescribed *time*.
Other methods of events are to replace with a value, or to multiply with a
value (see "events" help page of the package **deSolve**; ?events).

When the problem is solved, the existence of an event data.frame is specified by
passing to the solver, a list called events, which contains the data:

```
times <- seq(from = 0, to = 10, by = 1/24)
out2 <- ode(func = pharmaco2, times = times, y = yini,
            parms = NULL, method = "impAdams",
            events = list(data = injectevents))
```

The results (Fig. 3.8) show the instantaneous adjustment of the concentration in the
blood upon injection of the drug, and the exponential decrease in between injections.

```
plot(out2, lwd = 2, xlab="days")
```

### 3.4.2  A Bouncing Ball

In the previous pharmacokinetic examples, it was known in advance when a certain
event was occurring. This allowed us to specify the events in a data.frame
(Sect. 3.4.1.2), or to estimate the occurrence based on the simulation time (using
the modulo function in Sect. 3.4.1.1). The events either consisted of a change in the
problem specification (the derivative), when inputing the drug in the intestinal tract
(Sect. 3.4.1.1), or in a change in the value of the state variables when injecting the
drug directly (Sect. 3.4.1.2).

In many cases, we do not know in advance when a certain switch will occur, and
locating this will be part of the solution.

Consider the example of a bouncing ball [25], specified by its position above
the ground ($y$). The ball is thrown vertically, from the ground ($y(0) = 0$), with

initial velocity $y'$ of $10 \text{m s}^{-1}$. As the ball hits the ground, it bounces. This causes a sudden change in the value of the ball's velocity (a sign-reversal and reduction of its magnitude).

The differential equation and initial conditions specifying an object falling without friction through the air are:

$$\begin{aligned} y'' &= -g \\ y(0) &= 0 \\ y'(0) &= 10, \end{aligned} \tag{3.12}$$

where $y''$ is the acceleration, $y'$ the velocity and $y$ the height of the object above the ground.

Before this second order equation can be solved, it is rewritten as two first order equations, by including a description of the ball's velocity ($y_2 = y'_1$). The acceleration $g$ is taken as $9.8 \text{m s}^{-2}$.

$$\begin{aligned} y'_1 &= y_2 \\ y'_2 &= -9.8 \\ y_1(0) &= 0 \\ y_2(0) &= 10. \end{aligned} \tag{3.13}$$

Function `ball` specifies the differential system, which applies in between bounces. The dependent variables are the `height` ($y_1$) and `velocity` ($y_2$) of the ball.

```
library(deSolve)
yini   <- c(height = 0, velocity = 10)
```

```
ball <- function(t, y, parms) {
    dy1 <- y[2]
    dy2 <- -9.8

    list(c(dy1, dy2))
}
```

The ball bounce event is triggered by a `root` function, which signals when the ball hits the ground, i.e. when `y[1] = 0`. The root function thus simply returns `y[1]`:

```
rootfunc <- function(t, y, parms) y[1]
```

During the ball bounce (the "event"), its velocity (`y[2]`) is reversed and reduced by 10%. The event function must return both state variables:

```
eventfunc <- function(t, y, parms) {
    y[1] <- 0
    y[2] <- -0.9*y[2]
    return(y)
}
```

**Fig. 3.9**  The bouncing ball
model, including an event,
triggered by a root function.
See text for the R code

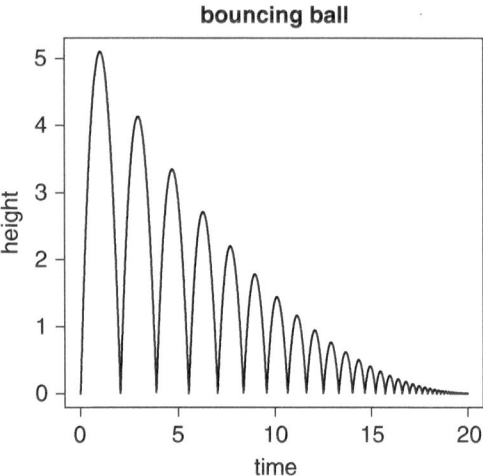

After specifying the output `times`, the model is solved. Several integration routines
in the package **deSolve** can locate the root of a function (see Table A.8). Here we
use the default method from `ode`, which is `lsoda`.

The solver needs to be informed that the event is triggered by a root (`root =
TRUE`), and the name of the event function (`func = eventfunc`) and the name
of the root function (`rootfun = rootfunc`) passed.

```
times <- seq(from = 0, to = 20, by = 0.01)
out <- ode(times = times, y = yini, func = ball,
           parms = NULL, rootfun = rootfunc,
           events = list(func = eventfunc, root = TRUE))
```

Fig. 3.9 shows how the ball bounces, each time loosing momentum:

```
plot(out, which = "height", lwd = 2,
     main = "bouncing ball", ylab = "height")
```

Note that the integrators detect the presence of an event in an integration step by a
sign change of the root function value. Therefore, if the function has multiple roots
in one step, some may be missed. Also, the solver will not be able to detect a root
that does not cross the zero value. For the bouncing ball example, this becomes clear
if we solve the model for a longer period; at a certain point in time, a root will go
unnoticed, and the ball's height will be negative and will be decreasing strongly in
time.

### 3.4.3   Temperature in a Climate-Controlled Room

We now deal with an example where the model dynamics change in response to a
certain *switching* function, i.e.

$$y' = f_1(x) \quad \text{if } g(x) = 1$$
$$y' = f_2(x) \quad \text{if } g(x) = 0, \tag{3.14}$$

where $g(x)$ is a *switch*.

Assume that in a climate-controlled room, the heat is switched on when the temperature drops below 18°C, and off when the room becomes warmer than 20°C. When the heating is on, the room warms at a constant rate, while there is constant cooling otherwise. The differential equations are thus different during the cooling and warming phases.

The challenge here is to know in which phase the system is. We could use a parameter, denoting the cooling or warming phase, but parameters cannot readily be changed (they are, by definition, assumed "constant" during the simulation).

Instead the problem is reformulated by adding the switching parameter as a state variable to the system of differential equations. The derivative of this state variable is set to 0 (i.e. it does not change during the integration), and its value is altered only when an event takes place, i.e. when the temperature exceeds a critical value and the heater turns on or off. Here is how we implement this model in R:

The model describes two state variables, the room temperature and the switching variable. The simulation starts off with a room temperature = 18°C, and heating switched on (y[2]=1).

The temperature (y[1]) either increases at a rate of 1°C per time unit, or decreases at a rate of $0.5°C t^{-1}$, depending on whether the switch state variable (y[2]) has a value 1 or 0. The derivative of the switch state variable, dy2 is 0:

```
yini   <- c(temp = 18, heating_on = 1)
```

```
temp <- function(t, y, parms) {
    dy1 <- ifelse(y[2] == 1, 1.0, -0.5)
    dy2 <- 0
    list(c(dy1, dy2))
}
```

The event is triggered when the temperature (y[1]) either takes the value of 18 or 20; this is at the root of either one of the functions y[1]-18 or y[1]-20:

```
rootfunc <- function(t, y, parms) c(y[1]-18, y[1]-20)
```

The event will switch the heating on or off, i.e. the switch will change from TRUE to FALSE or vice versa. Using the "!" (or "not") function is the simplest way to achieve this switch; note that the event function must return both state variables.

```
eventfunc <- function(t, y, parms) {
    y[1] <- y[1]
    y[2] <- ! y[2]
    return(y)
}
```

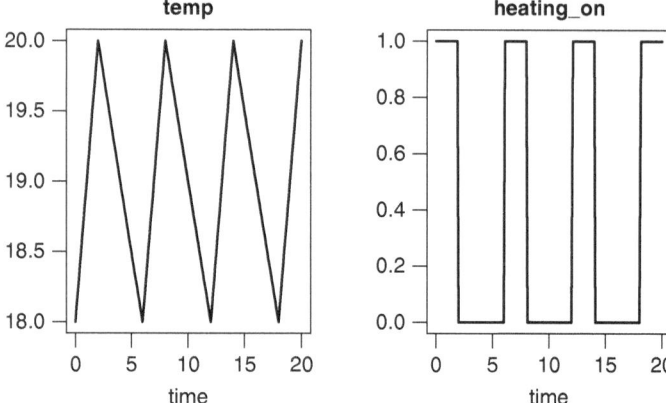

**Fig. 3.10** The temperature model, including an event, triggered by a root function. See text for the R code

The model is now solved with lsode, and the two state variables plotted versus time (Fig. 3.10)

```
times <- seq(from = 0, to = 20, by = 0.1)
out <- lsode(times = times, y = yini, func = temp,
             parms = NULL, rootfun = rootfunc,
             events = list(func = eventfunc, root = TRUE))
```

```
plot(out, lwd = 2)
```

The solver has stored the times at which the heating was turned on or off, in attribute troot:

```
attributes(out)$troot
```

```
[1]   2   6   6   6   8  12  14  18
```

You will notice that the same root at $t = 6$ was located three times. This is a numerical artifact. Even worse, if we extend the integration, the solver will at a certain point miss a root. Practical experience indicates that the root finding procedure that we implemented in radau is more robust for this type of model than the solver lsode that we used here. Solver radau will be discussed in Chap. 5.

## 3.5 Method Selection

If the problem is non-stiff we will normally use a predictor-corrector method or an explicit Runge-Kutta method, while for stiff problems, we will use an implicit method. Implicit methods require that, at each time step, a system of equations is solved to give the required solution (see Sect. 2.6). If the problem is stiff and linear then the algebraic equations to be solved are linear. In contrast, nonlinear equations are typically solved iteratively, using a variant of Newton's method [20]. This leads to a linear algebraic problem involving the Jacobian matrix at each iteration, and therefore multiple function evaluations per time step. Although in practice, the Jacobian will not be inverted (this is very inefficient), but rather Gaussian elimination will be used, this still requires quite a lot of computational overhead. On the other hand, as these multistep methods use previously computed values to evaluate the values at the next time step, they can attain high order of accuracy in less steps than taken by explicit methods, such as Runge-Kutta methods. The trade-off between number of steps and number of function evaluations per step, versus the overhead induced by the calculations involving the Jacobian determine which method is most efficient for a particular problem.

It will soon become very clear if we choose the wrong method. For instance, if an explicit method is chosen to solve a stiff problem, very small time steps will be taken in order to maintain stability, and it will take a long time to solve the problem. In such cases, the implicit bdf methods will be able to take much larger steps and solve the problem in a fraction of the time. In contrast, for non-stiff methods, the computational burden of Jacobian evaluation may overwhelm the fewer function evaluations needed. This is especially the case for very large sets of equations (e.g. resulting from numerically approximating partial differential equations, see Chap. 8), which, if they do not generate a stiff problem, may be much more efficiently solved with an explicit method.

It is generally not clear in advance which method may be best suited for a particular problem, but as using the optimal method may significantly improve overall performance, we give some rules of thumb to aid in the selection of the most appropriate method:

1. Use an implicit method only if the ODE problem is stiff; bdf, radau, mebdfi, gamd or bimd are best suited for very stiff problems, the adams methods for mildly stiff problems. The latter may also be more efficient for non-stiff problems, although the explicit Runge-Kutta methods are contenders in these cases.
2. If it is not known whether a problem is stiff, then use lsoda from the package **deSolve** or dopri5, cashkarp or dopri853 from the package **deTestSet** to print when problems become stiff. To provoke the printing of these features, set argument verbose=TRUE.
3. The diagnostics of a solution generated by the solvers provide a user with information about the number of function evaluations, and, for implicit methods, of the number of Jacobian decompositions. The diagnostics of method lsoda will tell the user which method was used, and when lsoda has switched between methods during the simulation.
4. Performance can be readily assessed by timing a model solution, using R's function system.time() method. So, as a crude approach, we can try several methods and simply take the one that requires least simulation time with a similar

accuracy (work precision diagrams described in Sect. 3.5.1.3 are very useful for this purpose).
5. We can also assess performance by recording the number of function or Jacobian evaluations.

### 3.5.1  The van der Pol Equation

A commonly used example to demonstrate stiffness is the van der Pol problem (see [11]). It is defined by the following second order differential equation:

$$y'' - \mu(1 - y^2)y' + y = 0, \tag{3.15}$$

where $\mu$ is a parameter. We convert (3.15) in a first order system of ODEs by adding an extra variable, representing the first order derivative:

$$\begin{aligned} y_1' &= y_2 \\ y_2' &= \mu(1 - y_1^2)y_2 - y_1. \end{aligned} \tag{3.16}$$

Stiff problems are obtained for large $\mu$, non-stiff for small $\mu$; the problems have both stiff and non-stiff parts for intermediate values of the parameter. We run the model for $\mu = 1, 10, 1000$, and using ode as the integrator:

```
yini   <- c(y = 2, dy = 0)
Vdpol <- function(t, y, mu)
     list(c(y[2],
             mu * (1 - y[1]^2) * y[2] - y[1]))
```

```
times <- seq(from = 0, to = 30, by = 0.01)
nonstiff <- ode(func = Vdpol, parms = 1,     y = yini,
                times = times, verbose = TRUE)
interm   <- ode(func = Vdpol, parms = 10,    y = yini,
                times = times, verbose = TRUE)
stiff    <- ode(func = Vdpol, parms = 1000, y = yini,
                times =0:2000, verbose = TRUE)
```

#### 3.5.1.1   Printing the Diagnostics of the Solutions

Function diagnostics prints the characteristics of the solutions. If we do this for the run with $\mu = 1$, then we see that lsoda solves the problem with the adams method, which requires 6009 function evaluations (Fig. 3.11a). In contrast, for $\mu = 10$, lsoda switches back and forth three times between the backward differentiation and the Adams formula, in the interval [0,30] (3897 function evaluations). The functions cashkarp and dopri5 consider the problem to become stiff at $t = 0.96$ and $t = 0.14$ respectively, while dopri853 does not

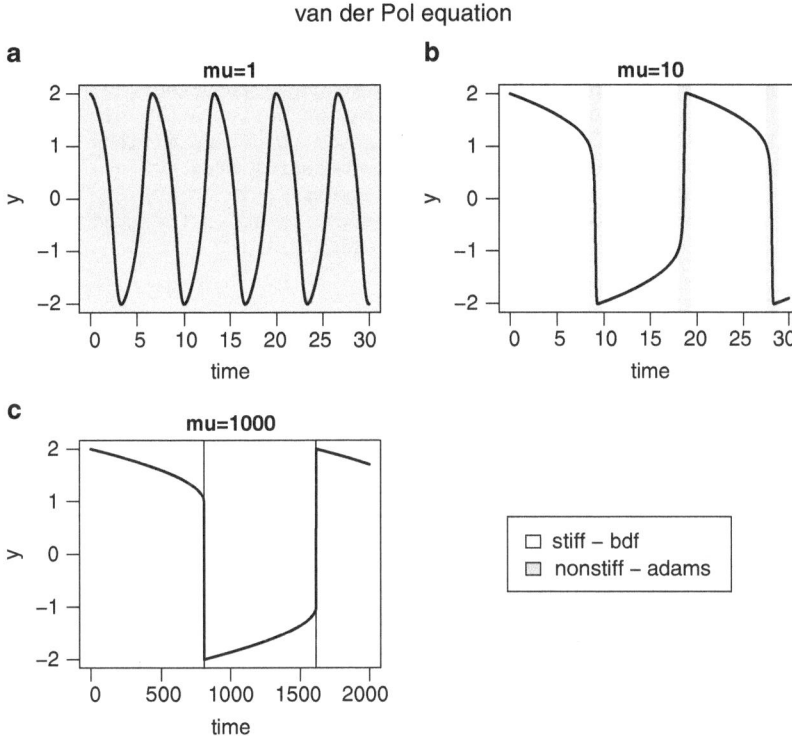

**Fig. 3.11** Three solutions of the van der Pol equation, solved with method lsoda. The regions where the solver uses the adams or the backward differentiation formula are indicated

consider this problem to be stiff. Method switching by lsoda also occurs twice when $\mu = 1000$, in the time interval [0, 2000] but here the region where the non-stiff method is used is very narrow (the thin grey line in Fig. 3.11c).

```
diagnostics(nonstiff)
```

```
--------------------
lsoda return code
--------------------

  return code (idid) =  2
  Integration was successful.


--------------------
INTEGER values
--------------------

  1 The return code : 2
  2 The number of steps taken for the problem so far: 3004
  3 The number of function evaluations for the problem so far: 6009
```

 5 The method order last used (successfully): 7
 6 The order of the method to be attempted on the next step: 7
 7 If return flag =-4,-5: the largest component in error vector 0
 8 The length of the real work array actually required: 52
 9 The length of the integer work array actually required: 22
14 The number of Jacobian evaluations and LU decompositions so far: 0
15 The method indicator for the last succesful step,
          1=adams (nonstiff), 2= bdf (stiff): 1
16 The current method indicator to be attempted on the next step,
          1=adams (nonstiff), 2= bdf (stiff): 1


--------------------
RSTATE values
--------------------

 1 The step size in t last used (successfully): 0.01
 2 The step size to be attempted on the next step: 0.01
 3 The current value of the independent variable which the solver has
   reached: 30.00947
 4 Tolerance scale factor > 1.0 computed when requesting too much
   accuracy: 0
 5 The value of t at the time of the last method switch, if any: 0


### 3.5.1.2  Timings

We can also run the same model with different integrators, each time printing the
time it takes (in seconds) to solve the problem:

```
library(deTestSet)
system.time(ode(func = Vdpol, parms = 10, y = yini,
                times = times, method = "ode45") )
```

```
   user   system elapsed
   0.29    0.00    0.30
```

```
system.time(ode(func = Vdpol, parms = 10, y = yini,
                times = times, method = "adams"))
```

```
   user   system elapsed
   0.06    0.00    0.06
```

```
system.time(ode(func = Vdpol, parms = 10, y = yini,
                times = times, method = "bdf"))
```

```
   user   system elapsed
   0.06    0.00    0.07
```

```
system.time(radau(func = Vdpol, parms = 10, y = yini,
                  times = times))
```

```
   user   system elapsed
   0.27    0.00    0.27
```

```
system.time(bimd(func = Vdpol, parms = 10, y = yini,
                 times = times))
```

```
 user   system elapsed
 0.17    0.00    0.18
```

```
system.time(mebdfi(func = Vdpol, parms = 10, y = yini,
                   times = times))
```

```
 user   system elapsed
 0.06    0.00    0.06
```

We ran the van der Pol problem with a number of different integration methods, each time using `diagnostics` to write the number of steps, function evaluations, Jacobian matrix decompositions and method switches performed by the integrators. Results are in Table 3.1. Clearly, the `adams` method which is most efficient for solving the non-stiff problem (requires fewest function evaluations), becomes completely unsuited in the stiff case, requiring more than eight million function evaluations! The implicit methods (`lsoda`, `bdf`, `impAdams`, `mebdfi`) perform rather well in all cases.

**Table 3.1** Performance of various integration routines implemented in **deSolve**, based on the van der Pol equation with different values of parameter $\mu$

| Method | Steps | Function evaluations | Jacobian evaluations | Switches to adams |
|---|---|---|---|---|
| times=seq(0,30,0.1) | $\mu = 1$ | | | |
| lsoda | 528 | 1173 | 0 | 0 |
| bdf | 789 | 1070 | 64 | |
| adams | 675 | 744 | 0 | |
| impAdams | 539 | 811 | 55 | |
| rk45ck | 300 | 1802 | 0 | |
| mebdfi | 659 | 2439 | 71 | |
| times=seq(0,30,0.1) | $\mu = 10$ | | | |
| lsoda | 705 | 1286 | 22 | 3 |
| bdf | 781 | 1096 | 71 | |
| adams | 1384 | 1681 | 0 | |
| impAdams | 625 | 896 | 60 | |
| rk45ck | 415 | 2492 | 0 | |
| mebdfi | 610 | 2383 | 73 | |
| times=0:2000 | $\mu = 1,000$ | | | |
| lsoda | 2658 | 3561 | 157 | 2 |
| bdf | 2744 | 3424 | 194 | |
| adams | 6829877 | 8730223 | 0 | |
| impAdams | 2646 | 3447 | 210 | |
| rk45ck | 1307209 | 7843256 | 0 | |
| mebdfi | 815 | 3200 | 101 | |

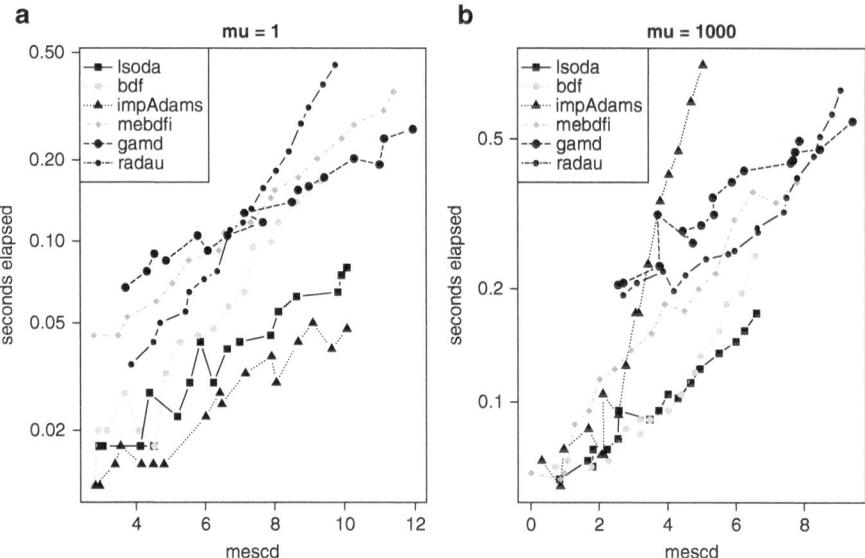

**Fig. 3.12** Work precision diagrams for the van der Pol problem

### 3.5.1.3   Work Precision Diagrams and mescd

In the previous section we compared the execution time of the codes, using the default relative and absolute tolerances. Most codes implement an error estimate (see Sect. 2.1.2.2), but it is not assured that the error will be of the same order as the prescribed tolerances. A common way to compare codes is to use the so-called work precision diagrams using the *mixed error significant digits*, *mescd*, defined by:

$$mescd := -\log_{10}(\max(|absolute\ error/(atol/rtol + |ytrue|)|)), \qquad (3.17)$$

where the *absolute error* is computed at all the mesh points at which output is wanted, *atol* and *rtol* are the input absolute and relative tolerances, *ytrue* is a more accurate solution computed using the same solver with smaller relative and absolute input tolerances and where (/, + and max) are element by element operators.

For every solver, a range of input tolerances were used to produce plots of the resulting mescd values against the number of CPU seconds needed for a run. Here we took the average of the elapsed CPU times of four runs. The format of these diagrams is as in ([11, 12], pp. 166–167, 324–325). As an example we report in Fig. 3.12 the work precision diagrams for the methods lsoda, bdf, impAdams, mebdfi, radau, gamd running the van der Pol problem with $\mu = 1$ and for $\mu = 1000$. The range of tolerances used is, for all codes, $rtol = 10^{-4+m/4}$ with $m = 0, 2, \ldots, 32$, all the other parameters are the default. We use times <- 0:30 for $\mu = 1$ and for $\mu = 1000$ we use times <-0:2000 for Fig. 3.12.

We want to emphasize that the reader should be careful when using these diagrams for a mutual comparison of the solvers. The diagrams just show the result of runs with the prescribed input on the specified computer. A more sophisticated setting of the input parameters, another computer or compiler, as well as another range of tolerances, or even another choice of the input vector times may change the diagrams considerably (not shown). For the van der Pol problem for $\mu = 1$ the impAdams is the most efficient code (Fig. 3.12a), while for $\mu = 1000$ lsoda and bdf require the least computational time to compute a solution with a similar number of *mescd* (Fig. 3.12b).

## 3.6 Exercises

### 3.6.1 Getting Started with IVP

Solve the problem

$$y' = y^2 + t, \tag{3.18}$$

with initial condition $y(0) = 0.1$ on the interval $[0,1]$; write the output to matrix out.
Now solve the following equations

$$y' = y^2 - yt, \tag{3.19}$$

and

$$y' = y^2 + 1, \tag{3.20}$$

with the same initial condition, and same output times. Save the output of the problems (3.19) and (3.20) to matrices out2, and out3 respectively. Plot the output of the three models in one plot.
Solve the following second order equation for $t \in [0,20]$.

$$y'' = -0.1y. \tag{3.21}$$

The initial conditions are $y(0) = 1$, $y'(0) = 0$. You will first need to rewrite this equation as two first order equations. Finally, solve the following differential problem using ode45

$$y'' + 2y' + 3y = \cos(t)$$
$$y(0) = y'(0) = 0, \tag{3.22}$$

in the interval $[0, 2\pi]$.

## 3.6.2   The Robertson Problem

This is a stiff problem consisting of three ordinary differential equations. It describes
the kinetics of an autocatalytic reaction given by [22]. The equations are:

$$y_1' = -k_1 y_1 + k_3 y_2 y_3$$
$$y_2' = k_1 y_1 - k_2 y_2^2 - k_3 y_2 y_3$$
$$y_3' = k_2 y_2^2. \tag{3.23}$$

Solve the problem in R; use as initial conditions $y_1 = 1$, $y_2 = 0$, $y_3 = 0$. The values
for the parameters are $k_1 = 0.04$; $k_2 = 3.10^7$; $k_3 = 1.10^4$. First integrate the problem
on the interval $0 \leq t \leq 40$. Then integrate it in the interval $10^{-4} \leq t \leq 10^7$.
   Use for the second output times a logarithmic series:

```
times <- 10^(seq(from = -4, to = 7, by = 0.1))
```

When plotting the outcome, scale the x-axis logarithmically.

## 3.6.3   Displaying Results in a Phase-Plane Graph

In (Sect. 3.2.1) the results of a three-equation model, the rigid body model, were
displayed in a 3D phase plane, using the R package **scatterplot3D**.

### 3.6.3.1   The Rossler Equations

Produce a 3-D phase-plane plot of the following set of ODEs, which you solve on
the interval [0, 100] and with initial conditions equal to (1, 1, 1):

$$y_1' = -y_2 - y_3$$
$$y_2' = y_1 + a y_2$$
$$y_3' = b + y_3(y_1 - c), \tag{3.24}$$

for $a = 0.2$, $b = 0.2$, $c = 5$. This system, called the Rossler equations, is due to [23];
its output is in Fig. 3.13, left (use `?scatterplot3d` to find out how to get rid of
the axis and grid).

### 3.6.3.2   Josephson Junctions

The next example, again from [12] describes superconducting Josephson Junctions.
The equations are:

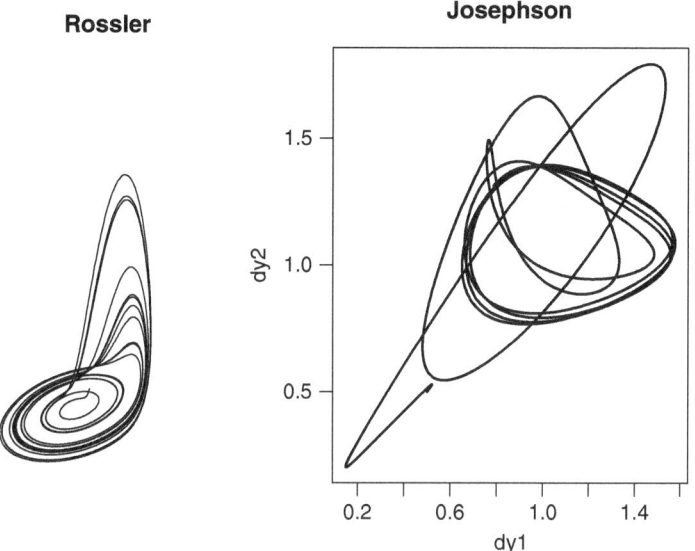

**Fig. 3.13** Phase plane of the Rossler equation (*left*) and the Josephson Junction (*right*)

$$c(y_1'' - \alpha y_2'') = i_1 - \sin(y_1) - y_1'$$
$$c(y_2'' - \alpha y_1'') = i_2 - \sin(y_2) - y_2'$$
$$y_1(0) = y_2(0) = y_1'(0) = y_2'(0) = 0.5, \tag{3.25}$$

Solve the equations in the interval $[0, 2\pi]$ and for $c = 2$, $\alpha = 0.5$, $i_1 = 1.11$, $i_2 = 1.18$.

You will need to write the equations as a function of $y_1''$ and $y_2''$ first, by taking suitable linear combinations. Then each second order differential equation should be rewritten as a set of two first order equations. For instance, solving for $y_1''$, we first rewrite:

$$cy_1'' - c\alpha y_2'' = i_1 - \sin(y_1) - y_1'$$
$$c\alpha y_2'' - c\alpha^2 y_1'' = \alpha(i_2 - \sin(y_2) - y_2'). \tag{3.26}$$

Then take the sum to obtain

$$c(1 - \alpha^2)y_1'' = i_1 - \sin(y_1) - y_1' + \alpha(i_2 - \sin(y_2) - y_2'). \tag{3.27}$$

Similarly, you can solve for $y_2''$.

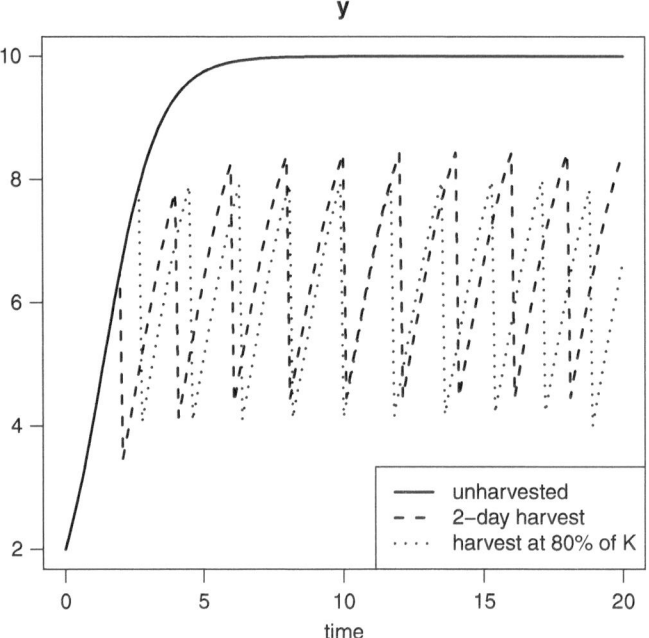

**Fig. 3.14**  Harvesting in the logistic model

### 3.6.4   Events and Roots

The logistic equation of Sect. 3.1.1, describes the growth of a population:

$$y' = ry\left(1 - \frac{y}{K}\right),\tag{3.28}$$

with $r = 1$, $K = 10$ and $y(0) = 2$.

Now this population is being harvested according to two strategies: one is to reduce the population's density every 2 days to 50%; the other is to wait until the species has reached 80% of its carrying capacity and then halving the density.

Implement these two strategies in a model. In the first, it will be easiest to outline the times at which harvesting occurs in a `data.frame`, much as we did in Sect. 3.4.1.2. In the second case you will need to use a root function that keeps track of when the population exceeds the critical density, and then reduce the density to 50% (see Sect. 3.4.2).

Run the model for 20 days and using three different scenarios. The first is when the population is unharvested, and the other two runs using each of the two strategies described above. Try to reproduce Fig. 3.14. You may also find inspiration on how to do this in the examples from **deSolve**'s events help page (`?events` will open this).

## 3.6.5  Stiff Problems

Several stiff test problems are described in [27]. One of their problems, called `dslode`, is given by:

$$y' = -\sigma(y^3 - 1), \tag{3.29}$$

where $\sigma$ is a parameter that determines the stiffness of the problem. Solve the problem in the interval $[0, 10]$, with $y(0) = 1.2$ and for three values of $\sigma$, equal to $10^6$, 1 and $10^{-1}$. Plot the three outputs in the same figure. Use function `diagnostics` to see how the integration was done (number of steps, method selected, etc...).

# References

1. Arenstorf, R. F. (1963). Periodic solutions of the restricted three-body problem representing analytic continuations of Keplerian elliptic motions. *American Journal of Mathematics, 85*, 27–35.
2. Aris, R. (1965). *Introduction to the analysis of chemical reactors*. Englewood Cliffs: Prentice Hall.
3. Bogacki, P., & Shampine, L. F. (1989). A 3(2) pair of Runge–Kutta formulas. *Applied Mathematics Letters, 2*, 1–9.
4. Brown, P. N., Byrne, G. D., & Hindmarsh, A. C. (1989). **VODE**, a variable-coefficient ODE solver. *SIAM Journal on Scientific and Statistical Computing, 10*, 1038–1051.
5. Brugnano, L., & Magherini, C. (2004). The BiM code for the numerical solution of ODEs. *Journal of Computational and Applied Mathematics, 164–165*, 145–158.
6. Cash, J. R., & Considine, S. (1992). An **MEBDF** code for stiff initial value problems. *ACM Transactions on Mathematical Software, 18*(2), 142–158.
7. Cash, J. R., & Karp, A. H. (1990). A variable order Runge–Kutta method for initial value problems with rapidly varying right-hand sides. *ACM Transactions on Mathematical Software, 16*, 201–222.
8. Dormand, J. R., & Prince, P. J. (1980). A family of embedded Runge–Kutta formulae. *Journal of Computational and Applied Mathematics, 6*, 19–26.
9. Dormand, J. R., & Prince, P. J. (1981). High order embedded Runge–Kutta formulae. *Journal of Computational and Applied Mathematics, 7*, 67–75.
10. Fehlberg, E. (1967). Klassische Runge–Kutta formeln funfter and siebenter ordnung mit schrittweiten-kontrolle. *Computing (Arch. Elektron. Rechnen), 4*, 93–106.
11. Hairer, E., & Wanner, G. (1996). *Solving ordinary differential equations II: Stiff and differential-algebraic problems*. Heidelberg: Springer.
12. Hairer, E., Norsett, S. P., & Wanner, G. (2009). *Solving ordinary differential equations I: Nonstiff problems* (2nd rev. ed.). Heidelberg: Springer.
13. Hindmarsh, A. C. (1980). **LSODE** and **LSODI**, two new initial value ordinary differential equation solvers. *ACM-SIGNUM Newsletter , 15*, 10–11.
14. Hindmarsh, A. C. (1983). **ODEPACK**, a systematized collection of ODE solvers. In R. Stepleman (Ed.), *Scientific computing: Vol. 1. IMACS transactions on scientific computation* (pp. 55–64). Amsterdam: IMACS/North-Holland.
15. Hundsdorfer, W., & Verwer, J. G. (2003). *Numerical solution of time-dependent advection-diffusion-reaction equations. Springer series in computational mathematics*. Berlin: Springer.

16. Iavernaro, F., & Mazzia, F. (1998). Solving ordinary differential equations by generalized Adams methods: Properties and implementation techniques. *Applied Numerical Mathematics, 28*(2–4), 107–126. Eighth conference on the numerical treatment of differential equations (Alexisbad, 1997).
17. Ligges, U., & Mächler, M. (2003). **Scatterplot3d**–an R package for visualizing multivariate data. *Journal of Statistical Software, 8*(11), 1–20.
18. Lorenz, E. N. (1963). Deterministic non-periodic flows. *Journal of Atmospheric Sciences, 20*, 130–141.
19. Petzold, L. R. (1983). Automatic selection of methods for solving stiff and nonstiff systems of ordinary differential equations. *SIAM Journal on Scientific and Statistical Computing, 4*, 136–148.
20. Press, W. H., Teukolsky, S. A., Vetterling, W. T., & Flannery, B. P. (2007). *Numerical recipes* (3rd ed). Cambridge: Cambridge University Press.
21. Reddy, M., Yang, R. S., Andersen, M. E., & Clewell, H. J., III (2005). *Physiologically based pharmacokinetic modeling: Science and applications*. Hoboken: Wiley.
22. Robertson, H. H. (1966). The solution of a set of reaction rate equations. In J. Walsh (Ed.), *Numerical analysis: An introduction* (pp. 178–182). London: Academic Press.
23. Rossler, O. E. (1976). An equation for continous chaos. *Physics Letters A, 57*(5), 397–398.
24. Shampine, L. F. (1994). *Numerical solution of ordinary differential equations*. New York: Chapman and Hall.
25. Shampine, L. F., Gladwell, I., & Thompson, S. (2003). *Solving ODEs with MATLAB*. Cambridge: Cambridge University Press.
26. Soetaert, K., Petzoldt, T., & Setzer, R. W. (2010). Solving differential equations in R: Package **deSolve**. *Journal of Statistical Software, 33*(9), 1–25.
27. van Dorsselaer, J. L. M., & Spijker, M. N. (1994). The error committed by stopping the newton iteration in the numerical solution of stiff initial value problems. *IMA journal of Numerical Analysis, 14*, 183–209.
28. Verhulst, P. (1838). Notice sur la loi que la population poursuit dans son accroissement. *Correspondance Mathematique et Physique, 10*, 113–121.

# Chapter 4
# Differential Algebraic Equations

**Abstract** In Chaps. 2 and 3 we were concerned mainly with the numerical solution
of ordinary differential equations of the form $y' = f(x,y)$. However, there are
problems which are more general than this and require special methods for their
solution. One such class of problems are differential algebraic equations (DAEs).
An important class of DAEs are those which can be written with a mass matrix $M$,
where the matrix $M$ is singular. Another important class of problems are differential
equations subjected to constraints. In this chapter we discuss solution methods
for these classes of equations. As we will see DAEs are often considerably more
difficult to solve than ODEs.

## 4.1 Introduction

In this book so far, we have been concerned with the numerical solution of (explicit)
first order systems of ODEs of the form:

$$y' = f(x,y). \tag{4.1}$$

There is another important class of problems which arise frequently in practice, and
which are more general than (4.1). These equations take the implicit form:

$$F(x,y,y') = 0. \tag{4.2}$$

If the matrix

$$\frac{\partial F}{\partial y'}(x,y(x),y'(x)), \tag{4.3}$$

is not singular, then, theoretically, any system of the form (4.2) can be converted
to the form (4.1), which can be solved by most of the initial value algorithms
considered earlier in this book. If the matrix (4.3) is singular then (4.2) are called
differential algebraic equations (DAEs).

K. Soetaert et al., *Solving Differential Equations in R*, Use R!,
DOI 10.1007/978-3-642-28070-2_4, © Springer-Verlag Berlin Heidelberg 2012

A commonly occurring special case of (4.2) is where a *mass* matrix ($M$) appears in the formulation of the problem, thus allowing (4.2) to be written in the more special form:

$$M(x,y)y' = f(x,y). \tag{4.4}$$

Problems described by (4.4) arise when there is a relationship that must hold, for all $x$, between some of the dependent variables. For example, in electrical circuit modelling, the equations ensure that summed currents and voltages passing through nodes and loops are zero. From problem formulation (4.4), it is immediately clear that, if the mass matrix is not singular (i.e. its inverse, $M^{-1}(x,y)$ exists), one way to solve this problem is to write it as:

$$y' = M^{-1}(x,y)f(x,y), \tag{4.5}$$

thus obtaining an ODE. In many practical applications however, the matrix $M(x,y)$ is singular and the numerical solution needs particular care because we must deal with the implicit equations related to the algebraic constraints. A typical example is given by the class of problems where the matrix $M$ is a constant diagonal matrix $M = diag(1,\ldots,1,0,\ldots,0)$, in this case the differential equations have the form

$$\begin{aligned} y' &= f(x,y,z) \\ 0 &= g(x,y,z). \end{aligned} \tag{4.6}$$

For example, in chemical reaction equations, the algebraic equations may enforce mass balance or represent so-called equilibrium expressions.

There are now several excellent books that deal with the theory of DAEs. The interested reader is advised to consult [1, 2, 7] and the references contained therein.

### 4.1.1   The Index of a DAE

A fundamentally important concept in both the theory and the design of algorithms for the numerical solution of DAEs is that of the *index* of a DAE. In a sense this tells us how far away the DAE is from being an ODE. The higher the index, the further it is from an ODE, and, as we might expect, the more difficult it is in general to solve the DAE. There exist many different types of DAE-indices. We mention the *differentiation index* [7, p. 454], the *perturbation index* [7, p. 459] and the *tractability index* [12].

Here we use the following definition [1]: for a general DAE of the form (4.2) the *differentiation index* along a solution $y(x)$ is the minimum number of differentiations of the DAE with respect to $x$ that are required to solve for $y'$ uniquely in terms of $y$ and $x$. Note in particular that the index depends on the solution [1, p. 236], [7, p. 454].

As an example, consider the equation

$$y_1' = y_2$$
$$y_1 = g(x).$$

(4.7)

Differentiating the algebraic equation with respect to $x$ once, we obtain:

$$y_1' = y_2 = g'(x).$$

(4.8)

We need another differentiation to obtain an ODE:

$$y_2' = g''(x),$$

(4.9)

so this equation has differentiation index $= 2$ because two differentiations of $g(x)$ were needed to obtain an ODE.

One way to solve DAEs is to rewrite them as ODEs by performing these analytic differentiations. This technique is known as index-reduction. However, this has some important drawbacks, and so this approach is not much used in practice. Finally we note that, of course, an ODE is a differential algebraic equation of index 0.

### 4.1.2 A Simple Example

As a first example, we consider the famous "Robertson" problem which describes an autocatalytic reaction [15] between three chemical species, $A$, $B$ and $C$:

$$A \xrightarrow{k_1} B$$
$$B + B \xrightarrow{k_2} C + B$$
$$B + C \xrightarrow{k_3} A + C,$$

(4.10)

where the numbers on top of the arrows are the reaction rate coefficients, generally taken to be 0.04, $3 \cdot 10^7$ and $10^4$ for $k_1, k_2, k_3$ respectively.

This problem is usually formulated as an ODE [13]:

$$A' = -k_1 A + k_3 BC, \qquad A(0) = 1$$
$$B' = k_1 A - k_3 BC - k_2 B^2, \qquad B(0) = 0$$
$$C' = k_2 B^2, \qquad C(0) = 0.$$

(4.11)

As the sum of all derivatives $(A' + B' + C')$ is 0, the total mass is constant (and equal to $A(0) + B(0) + C(0) = 1$), so an equivalent representation of the system, as a DAE is:

$$A' = -k_1 A + k_3 BC$$
$$B' = k_1 A - k_3 BC - k_2 B^2$$
$$1 = A + B + C,$$

(4.12)

where the first two equations are differential equations that specify the dynamics of chemical species $A$ and $B$, while the third algebraic equation ensures that the summed concentration of the three species remains at 1. Written in the form (4.2), this becomes:

$$
\begin{aligned}
0 &= -A' - k_1A + k_3BC \\
0 &= -B' + k_1A - k_3BC - k_2B^2 \\
0 &= A + B + C - 1,
\end{aligned}
\tag{4.13}
$$

while in the form (4.4):

$$
\begin{bmatrix} 1 & 0 & 0 \\ 0 & 1 & 0 \\ 0 & 0 & 0 \end{bmatrix} \cdot \begin{bmatrix} A' \\ B' \\ C' \end{bmatrix} = \begin{bmatrix} -k_1A + k_3BC \\ k_1A - k_3BC - k_2B^2 \\ A + B + C - 1 \end{bmatrix}.
\tag{4.14}
$$

Now the index of (4.12) is 1 as we need to differentiate the algebraic equation only once to obtain an ODE:

$$
\begin{aligned}
0 &= A' + B' + C' \\
C' &= -A' - B' \\
C' &= k_2B^2.
\end{aligned}
\tag{4.15}
$$

This is the original formulation (4.11).

### 4.1.3  DAEs in Hessenberg Form

An important class of DAEs for which there exist many theoretical results are DAEs in Hessenberg form see [1, p. 238]. An index 1 DAE in Hessenberg form is given by system (4.6). For simplicity we rewrite this here in autonomous form:

$$
y' = f(y, z)
\tag{4.16}
$$

$$
0 = g(y, z),
\tag{4.17}
$$

where $g_z$ [1] is assumed to be non singular for all $x$ in a neighbourhood of the solution. We note that a non-autonomous system can be written in the autonomous form by adding an equation for the independent variable $x$ i.e. $x' = 1$.

An index 2 DAE in Hessenberg form is given by the following equations:

$$
y' = f(y, z)
\tag{4.18}
$$

$$
0 = g(y),
\tag{4.19}
$$

---

[1] $g_z$ is shorthand for $\dfrac{\partial g}{\partial z}$.

where $g_y f_z$ is assumed to be non singular for all $x$ in a neighbourhood of the solution. Differentiating (4.19) shows that the solution has to satisfy the equation

$$0 = g_y(y) f(y, z), \tag{4.20}$$

and we can use this equation to determine the $z$ component in a locally unique way. Equation (4.20) is known as a hidden constraint.

An index 3 DAE in Hessenberg form is given by:

$$y' = f(y, z) \tag{4.21}$$

$$z' = k(y, z, u) \tag{4.22}$$

$$0 = g(y), \tag{4.23}$$

where $g_y f_z k_u$ is assumed to be non singular for all $x$ in a neighbourhood of the solution.

### 4.1.4   Hidden Constraints and the Initial Conditions

For initial value DAE problems not all components of the initial value can be chosen freely which is in contrast to the case for ODEs. For certain DAEs, it may not be obvious how to formulate the initial conditions that lead to a uniquely solvable IVP.

The initial conditions of a DAE are consistent only if they obey the algebraic equation. For systems of index 1 (as in the Robertson problem, Sect. 4.1.2, equation (4.12)), the initial conditions are straightforward to derive. However, for higher index systems, there can be hidden constraints. These are important as we need to take care of them to make the initial conditions consistent. The problem of finding consistent initial conditions has been analyzed by many authors (see [11] and the references therein). In what follows we will consider the concept of hidden constraints and show how these arise in the numerical integration of DAEs in Hessenberg form.

#### 4.1.4.1   Hidden Constraints in Index 2 DAEs in Hessenberg Form

We consider first of all the system of index 2 defined by (4.18) and (4.19). In this system, (4.20) defines the hidden constraint. Consequently, the initial conditions should satisfy the following equations:

$$\begin{aligned} g(y_0) &= 0 \\ g_y(y_0) f(y_0, z_0) &= 0, \end{aligned} \tag{4.24}$$

in order to be consistent.

**Fig. 4.1** Schematic
representation of the
pendulum problem

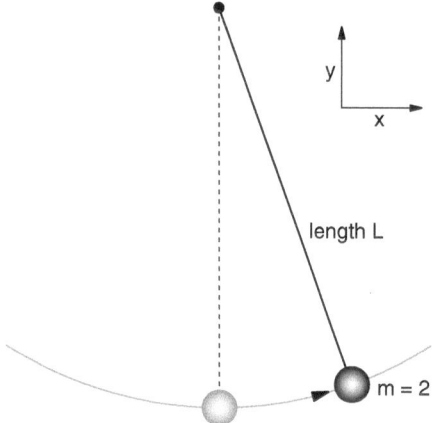

### 4.1.4.2   Hidden Constraints in Index 3 DAEs

Now we consider systems of index 3 defined by (4.21)–(4.23). Differentiating (4.23)
we have

$$0 = g_y(y)f(y,z)$$
$$0 = g_{yy}(f,f) + g_y f_y f + g_y f_z k.$$

(4.25)

Consistent initial values must satisfy (4.23) and the above two hidden conditions (4.25).

## 4.1.5   The Pendulum Problem

The classic example of an index 3 problem in Hessenberg form is the pendulum
problem which is a simple mechanical system [2, 7] (Fig. 4.1).

   The equations modeling the motion of a simple pendulum of length $L$ with an
infinitesimally small ball of mass $m$, tension $\lambda$ and only gravitational forces acting
on the pendulum, are given by two second order ODEs and one algebraic equation.
The latter results from a geometrical constraint on the variables $x$ and $y$.

$$mx'' = -2\lambda x$$
$$my'' = -2\lambda y - mg$$
$$L^2 = x^2 + y^2.$$

(4.26)

Here $x$ and $y$ are the position of the mass in cartesian coordinates and $g$ is gravity.
If we take $m = 2$, and we rewrite this system in first order form by adding two
additional variables, we obtain:

$$x' = u$$
$$y' = v$$
$$u' = -\lambda x \qquad (4.27)$$
$$v' = -\lambda y - g$$
$$0 = x^2 + y^2 - L^2,$$

where $u$ and $v$ are the ball's velocity in the $x$- and $y$-direction respectively. The dependent variables are $x, y, u, v$ and $\lambda$, where $\lambda$ is also called a Lagrange multiplier. Note that its dynamics are not described by an ODE.

## 4.2 Solving DAEs

We start this section with a note of caution. Our aim is to discuss numerical algorithms for the solution of the general DAE (4.2) written in implicit form. However (4.2) can include problems which are not well defined mathematically and which can not be solved by any method based on direct discretization of $y$ and $y'$ (see [1, p. 238]).

### 4.2.1 Semi-implicit DAEs of Index 1

One of the easiest classes of DAEs to analyse and indeed to solve is of the form:

$$y' = f(x, y, z) \qquad (4.28)$$
$$0 = g(x, y, z), \qquad (4.29)$$

where the differential and the algebraic equations are separated. This can be regarded as an ODE coupled with constraints. If the matrix $\frac{\partial g}{\partial z}$ is invertible in the neighborhood of the solution, then (4.28) and (4.29) is an index 1 DAE. The initial conditions are consistent if $0 = g(x_0, y_0, z_0)$.

There exist two commonly used approaches to the numerical solution of this type of DAE: the "$\varepsilon$-embedding method" and the "state space form method" [7].

#### 4.2.1.1 The $\varepsilon$-embedding Method

In the $\varepsilon$-embedding method for problems of index 1, we consider the ODE system:

$$y' = f(x, y, z)$$
$$\varepsilon z' = g(x, y, z). \qquad (4.30)$$

The behaviour of the solution of (4.30) is studied as $\varepsilon \to 0$. This will give us valuable information concerning the behaviour of our numerical problems for extremely stiff cases and also suggest algorithms for solving (4.28) and (4.29). For a description of the $\varepsilon$ embedding approach the reader is referred to [7, p. 374]. For a description of methods which are applicable to the differential equation (4.30) the reader is referred to Chaps. 1 and 2 earlier in this book.

#### 4.2.1.2   The State Space Form Method

In what follows, we will assume that the functions $y$ and $g$ are sufficiently often differentiable for our purposes and we will be concerned with the case where $\frac{\partial g}{\partial z}(x, y, z)$ is invertible in the neighborhood of the solution.

Under these assumptions the implicit function theorem guarantees that (4.29) possesses a locally unique solution $z = \bar{g}(x, y)$, which can be substituted in (4.28) to give the differential equation $y' = f(x, y, \bar{g}(x, y))$. This ordinary differential equation is the so-called state-space form. For instance, as $C = 1 - A - B$, it is easy to see that the Robertson problem (4.12) can be written in state space form as:

$$
\begin{aligned}
A' &= -k_1 A + k_3 B \cdot (1 - A - B) \\
B' &= k_1 A - k_3 B \cdot (1 - A - B) - k_2 B^2.
\end{aligned}
\tag{4.31}
$$

Often the algebraic equations (4.29) have a special structure, which can be used to solve them very efficiently, for instance if an analytic solution exists, as in the previous (Robertson) example. In this case, this solution method can be faster than using special DAE solvers, where the algebraic equations are solved with standard, iterative methods. However, as we will see later, there are various reasons for which this procedure is not always recommended.

### 4.2.2   General Implicit DAEs of Index 1

We now show how a general implicit equation of the form (4.2), can be solved using standard numerical methods. This approach was first proposed by [6] and [5], for use with Backward Differentiation Formulae (BDF). The idea is to take (4.2) and to replace the derivative $y'$ by a finite difference approximation. The most simple approach is to replace $y'$ with the backward Euler method to obtain

$$
F\left(x_n, y_n, \frac{y_n - y_{n-1}}{h}\right) = 0,
\tag{4.32}
$$

with $h = x_n - x_{n-1}$. This equation can be solved for $y_n$, thus advancing the solution from $(x_{n-1}, y_{n-1})$ to $(x_n, y_n)$. It is straightforward to show that for semi-explicit index 1 DAEs the Backward Euler method is first order accurate, stable and convergent.

An obvious way to extend this to higher order is to use higher order linear multistep or Runge–Kutta methods. We demonstrate this by using a higher order BDF. Following [5] we replace $y'_n$ in (4.2) with a $k$-step BDF so that

$$y'_n = \frac{1}{h\beta_k} \sum_{i=0}^{k} \alpha_{k-i} y_{n-i}, \qquad (4.33)$$

where $\alpha_{k-i}$ and $\beta_k$ are the coefficients of the BDF methods. This expression can be inserted into (4.2) to produce the system of nonlinear algebraic equations defining $y_n$:

$$F(x_n, y_n, y'_n) = 0. \qquad (4.34)$$

Now if the original (4.2) is of index 1 and differentiable with respect to both $y$ and $y'$, then its solution using a BDF with fixed $h$ and $k < 7$ converges with order $O(h^k)$ if all initial values are given correct to $O(h^k)$ and the Newton iteration scheme is solved to $O(h^{k+1})$.

One of the most powerful methods available for solving index 1 DAEs is the code DASSL [14] which is based on BDF. Another efficient approach to solving index 1 DAEs is via the use of Runge–Kutta methods and this is explained in detail in [7, p. 371]. Experience has shown that index 1 DAEs are often not much more difficult to solve than explicit ODEs written in the form (4.1).

### 4.2.3 Discretization Algorithms

We finish off this section with a discussion of how we might solve differential algebraic equations using discretization. This has been considered in some detail in [7, p. 490], but it is instructive to summarise it here. Our aim is to illustrate how Newton's method is applied to the discretization of a DAE system.

If we consider the Backward Euler method applied to:

$$\begin{aligned} y' &= f(y, z) \\ 0 &= g(y), \end{aligned} \qquad (4.35)$$

we have

$$\begin{aligned} y_{n+1} - y_n - h f(y_{n+1}, z_{n+1}) &= 0 \\ g(y_{n+1}) &= 0. \end{aligned} \qquad (4.36)$$

The jacobian matrix associated with (4.36) is:

$$J = \begin{pmatrix} I - h f_y & -h f_z \\ g_y & 0 \end{pmatrix}, \qquad (4.37)$$

where we now leave out the indices of the variables. We now use the simplified Newton iteration (a fixed point iteration)

$$0 = J \begin{pmatrix} y_{n+1} \\ z_{n+1} \end{pmatrix} - \begin{pmatrix} y_{n+1} - y_n - hf(y_{n+1}, z_{n+1}) \\ g(y_{n+1}) \end{pmatrix}. \tag{4.38}$$

A detailed analysis of this scheme can be found in [7, p. 491].

## 4.2.4   DAEs of Higher Index

Of course, not all DAEs are of index 1. In fact, many problems of practical importance have index > 1.

There are two fundamental choices that can be taken in order to solve higher index equations: either we use numerical methods suitable for dealing with high index problems or we reduce the index of the system (to 1 if possible).

In what follows we will illustrate the latter approach by means of the well-known pendulum problem.

### 4.2.4.1   Index Reduction

An obvious approach to reduce the index of a system is to differentiate it. We saw an example of this in Sect. 4.1.2. In theory we can differentiate until we obtain a set of ODEs, or we may stop at index 1 as these problems are not much more difficult to solve than ODEs (note that a problem of index 0 is a differential equation).

Following this approach we can reduce the index of the pendulum problem (4.27) by successive differentiation of the algebraic constraints.

Differentiating the algebraic equation (the last from (4.27)) once and simplifying, we obtain:

$$2xx' + 2yy' = 0$$

or                                                                          (4.39)

$$xu + yv = 0.$$

This reformulates the constraint and reduces the system's index to two.

The complete index two pendulum problem in Hessenberg form is:

$$\begin{aligned} x' &= u \\ y' &= v \\ u' &= -\lambda x \\ v' &= -\lambda y - g \\ 0 &= xu + yv. \end{aligned} \tag{4.40}$$

Differentiating the constraint once more, substituting for $u'$ and $v'$, and simplifying we obtain:

$$u^2 + v^2 + xu' + yv' = 0$$
$$u^2 + v^2 - (x^2 + y^2)\lambda - yg = 0 \qquad\qquad (4.41)$$

or

$$u^2 + v^2 - L^2\lambda - yg = 0,$$

where the system is now of index 1. The algebraic equation can be rewritten to provide an expression for $\lambda$ and this expression can be used to find a suitable initial condition for $\lambda$.

We now differentiate the equation once more and solve for $\lambda'$:

$$2uu' + 2vv' - L^2\lambda' - y'g = 0$$
$$\lambda' = \frac{1}{L^2}(-2\lambda(ux + vy) - 3vg). \qquad\qquad (4.42)$$

So, after three differentiations, we obtain an expression for the derivative of $\lambda$, and the algebraic equation has been converted to an ODE. Consequently, the original set of equations is an index 3 system.

It is very important to note that, except for the original index 3 formulation, none of the equations guarantees that $x^2 + y^2 = L^2$. This leads to the main problem of index reduction, namely that the error in the constraint will grow with each reduction. This is illustrated in Fig. 4.2 for the four different formulations of the pendulum problem. We see that, whereas the error stays small when solved in index 3 form (Fig. 4.2a), the error increases when differentiated once (Fig. 4.2b), but the increase is quadratic and of oscillatory nature when differentiated twice or more (Fig. 4.2c, d). This phenomenon is known as drift off.

The error in the original constraint and in the hidden constraints for the four problem formulations are given in Table 4.1. The problem occurs because the algebraic constraints are not very well preserved by the numerical solution. One way to avoid this instability is to repeatedly project the numerical solution onto all constraints. The interested reader is referred to [7, p. 470].

### 4.2.4.2  Higher Index Solvers

As mentioned previously, it is often much more difficult to solve DAEs than it is to solve ODEs. Indeed only index 1 DAEs can safely be solved numerically. However for specially structured DAEs there are numerical methods which may work well. Three very efficient general purpose codes for the solution of DAEs are BDF methods (typified by DASSL [14]), RADAU5 [7] and MEBDF (typified by MEBDFDAE and MEBDFI [4]). In particular the codes RADAU5 and MEBDF can deal with problems of index $\leq 3$ that are in Hessenberg form. The codes DASSL [14] and DASPK2.0 [2] can solve problems of index $\leq 1$, while DASPK3.0 [2] also solves and initialises DAEs of index 2 in Hessenberg form. The DASPK codes can solve very large problems such as those arising in the method of lines solution

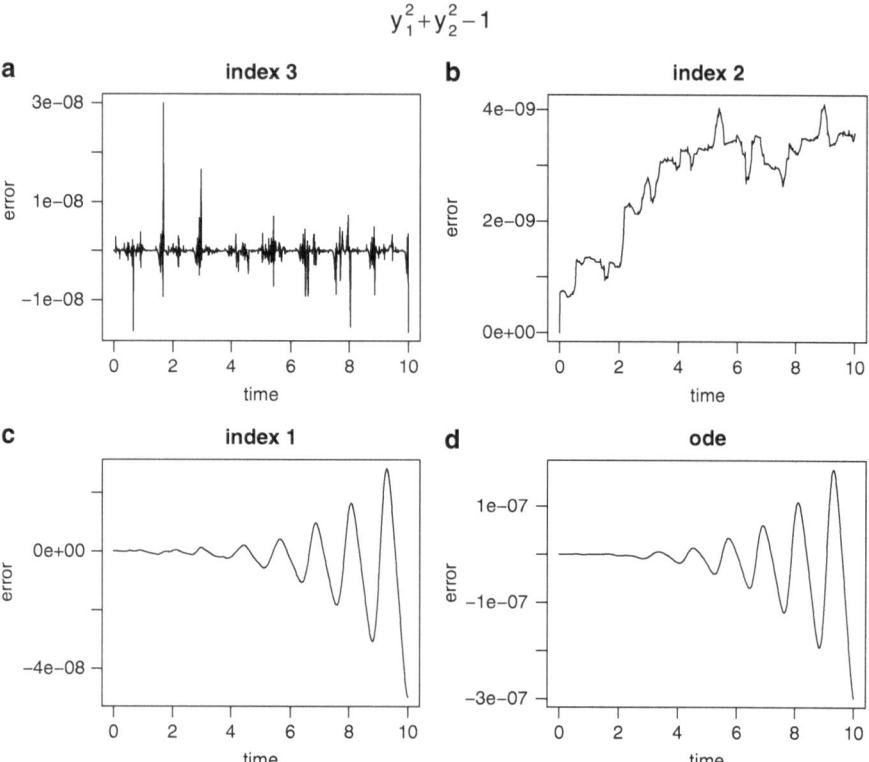

$$y_1^2 + y_2^2 - 1$$

**Fig. 4.2** When the index of the well-known pendulum problem, an index 3 system, is reduced by differentiation, the numerical solution will start to drift away from the algebraic constraint $L^2 = y_1^2 + y_2^2$, with $L = 1$. (**a**) This error is small in the original formulation. (**b**) The error increases with time for the index 2 formulation. (**c**, **d**) Worse, it oscillates with exponentially increasing amplitude for index 1 (**c**) and index 0 (**d**) systems. The solution was obtained with mebdfi, using tolerances $= 10^{-8}$

**Table 4.1** Mean of the absolute value of the constraint and the hidden constraints for numerical solutions computed with MEBDFI (tolerances $= 10^{-8}$) for the four formulations of the pendulum equation

| Index | $x^2 + y^2 - 1$ | $xu + yv$ | $u^2 + v^2 - \lambda - gy$ |
|-------|-----------------|-----------|----------------------------|
| 3 | $8.59 \cdot 10^{-10}$ | $3.74 \cdot 10^{-7}$ | $1.72 \cdot 10^{-4}$ |
| 2 | $2.74 \cdot 10^{-9}$ | $6.40 \cdot 10^{-10}$ | $5.02 \cdot 10^{-7}$ |
| 1 | $6.21 \cdot 10^{-9}$ | $1.56 \cdot 10^{-8}$ | $4.35 \cdot 10^{-9}$ |
| 0 | $3.78 \cdot 10^{-8}$ | $9.98 \cdot 10^{-8}$ | $6.10 \cdot 10^{-8}$ |

of PDEs (see Sect. 8.4). This efficiency is obtained by using Krylov subspaces to solve the linear algebraic equations. Two other codes that can handle problems with index $\leq 3$ that are in Hessenberg form are the codes GAMD [10] and BIMD [3].

It is important to make clear that it is often not possible for a code to solve non-Hessenberg (general) higher index systems routinely. This is shown in [2, p. 45] where a stable linear index 2 DAE which depends on a parameter can be unstable or unsolvable for all Runge–Kutta and linear multistep methods.

There is an assortment of other methods which can be very effective in the right circumstances. An approach which could be adopted for high index DAEs is first to reduce the index of the problem by differentiating it and then to try either a projection method, the state phase form or a half explicit method to solve the problem. The interested reader is referred to [7, Chap. 7].

### 4.2.5   Index of a DAE Variable

One drawback of solvers of higher index DAEs is that they need specification of the "index of each variable" separately. This notion "index of a variable" was introduced for the first time in the documentation of the code RADAU5 and it is strictly connected to the error estimate used in that code [7] and later used in the codes MEBDFDAE, MEBDFI, BIMD and GAMD.

An analysis of the local error estimate for Runge–Kutta methods for the component $z$ of (4.18) shows that the error grows like $O(1/h)$ when the step-size decreases [8, p. 102]. Since this is a dangerous property for a variable step-size implementation it is advisable to scale the $z$ component with a factor $h$. For this reason the $z$ components are called "index 2" variables.

A rigorous analysis has been made for index 3 DAEs in Hessenberg form [8]. For this class of problems we have that, for the component $z$ (in equation (4.21)), the error grows like $O(1/h)$, and for the component $u$ (equation (4.22)) the error grows like $O(1/h^2)$ when the step-size decreases. For this reason it is advisable to scale the $z$ component with a factor $h$ and the $u$ component with a factor $h^2$. Now, for obvious reasons, the $z$ components are called "index 2" variables and the $u$ components are called the "index 3" variables. The variables that do not require a scaling are called "index 1" variables.

The original pendulum problem (4.27) is an example of an index 3 problem in Hessenberg form. For this problem $(x,y)$ are the "index 1" variables, $(u,v)$ are the "index 2" variables and $\lambda$ is the "index 3" variable.

Based on (4.40), it is clear that for the index two pendulum problem $(x,y,u,v)$ are the "index 1" variables and $\lambda$ is the "index 2" variable. Finally, for the index 1 formulation of the pendulum problem (4.41), all variables are of index 1.

We observe that the index of a variable is clearly defined only for the class of problems written in Hessenberg form. In this case the index of the DAE equals the maximum index of a variable. For general DAEs of the form (4.2) it is, in general, not possible to associate an index to one variable.

# References

1. Ascher, U. M., & Petzold, L. R. (1998). *Computer methods for ordinary differential equations and differential-algebraic equations.* Philadelphia: SIAM.
2. Brenan, K. E., Campbell, S. L., & Petzold, L. R. (1996). *Numerical solution of initial-value problems in differential-algebraic equations.* Philadelphia: SIAM Classics in Applied Mathematics.
3. Brugnano, L., Magherini, C., & Mugnai, F. (2006). Blended implicit methods for the numerical solution of DAE problems. *Journal of Computational and Applied Mathematics, 189*(1–2), 34–50.
4. Cash, J. R., & Considine, S. (1992). An **MEBDF** code for stiff initial value problems. *ACM Transactions on Mathematical Software, 18*(2), 142–158.
5. Gear, C. W. (1990). Differential-algebraic equations, indices and integral algebraic equations. *SIAM Journal on Numerical Analysis, 27*, 1527–1534.
6. Gear, C. W., & Petzold, L. R. (1984). ODE methods for the solution of differential/algebraic systems. *SIAM Journal on Numerical Analysis, 21*, 716–728.
7. Hairer, E., & Wanner, G. (1996). *Solving ordinary differential equations II: Stiff and differential-algebraic problems.* Heidelberg: Springer.
8. Hairer, E., Lubich, C., & Roche, M. (1989) *The numerical solution of differential-algebraic systems by Runge-Kutta methods: Vol. 1409. Lecture notes in mathematics.* Berlin etc.: Springer. vii, 139 p. DM 25.00.
9. Hairer, E., Norsett, S. P., & Wanner, G. (2009). *Solving ordinary differential equations I: Nonstiff problems. second revised edition.* Heidelberg: Springer.
10. Iavernaro, F., & Mazzia, F. (1998). Solving ordinary differential equations by generalized Adams methods: Properties and implementation techniques. *Applied Numerical Mathematics, 28*(2–4), 107–126. Eighth conference on the numerical treatment of differential equations (Alexisbad, 1997).
11. Lamour, R., & Mazzia, F. (2009). Computation of consistent initial values for properly stated index 3 DAEs. *BIT, 49*(1), 161–175.
12. März, R. (2002). The index of linear differential algebraic equations with properly stated leading terms. *Results in Mathematics, 42*(3–4), 308–338.
13. Mazzia, F., & Magherini, C. (2008). *Test set for initial value problem solvers, release 2.4* (Rep. 4/2008). Department of Mathematics, University of Bari, Italy.
14. Petzold, L. R. (1983). A description of **DASSL**: A differential/algebraic system solver. *IMACS trans on Scientific Computation*, New Brunswick, NJ, 65–68.
15. Robertson, H. H. (1966). The solution of a set of reaction rate equations. In J. Walsh (Ed.), *Numerical analysis: An introduction* (pp. 178–182). London: Academic.

# Chapter 5
# Solving Differential Algebraic Equations in R

**Abstract**  R contains several methods for the solution of initial value problems for
DAEs, which are embedded in the R packages **deSolve** and **deTestset**. Four of these,
based on RADAU5, MEBDF, block implicit or Adams methods, can solve DAEs of
index up to three written in Hessenberg form. The fifth method, based on BDF, is
very efficient for index 1 problems and can solve some higher index problems as
well. We illustrate how to solve DAEs as they arise in the modelling of constrained
mechanical systems, electrical circuits, and chemical (equilibrium) reactions.

## 5.1  Differential Algebraic Equation Solvers in R

There are currently five R functions that solve inital value problems for differential
algebraic equations.

Two methods, `daspk` [2] from the R package **deSolve** [12] and `mebdfi` [4]
from the R package **deTestSet** [13] can solve DAEs written in the general implicit
form:

$$r(t, y, y', p) = 0, \tag{5.1}$$

where $r$ is the residual function and $p$ are the parameters. Function `daspk` [2]
implements a backward differentiation formula (BDF), related to DASSL [9],
which is particularly effective in solving DAEs of index up to 1,[1] while `mebdfi`
[4] implements the modified extended backward differentiation formulae and can
handle problems of index up to 3 which are in Hessenberg form.

The function `gamd` [7] and `bimd` [3] from **deTestSet** are based on generalised
Adams methods and blended implicit methods respectively, while another method,
the **deSolve** function `radau` [6] is based on a three-stage implicit Runge-Kutta

---

[1]Note that in the R implementation of daspk, it is possible to scale the higher index variables
as described in Sect. 4.2.5. Therefore, the R function `daspk` can also solve certain higher index
problems.

formula. These functions can solve linearly implicit DAEs up to index 3, which are
written in the form

$$My' = f(t, y, p).$$ (5.2)

A simplified form of the syntax for solving DAEs, written in implicit form, in R is:

```
daspk(y, times, parms, dy, res, ...)
mebdfi(y, times, parms, dy, res, nind, ...)
```

where `times` holds the times at which output is wanted; `y` holds the initial
conditions, `dy` the initial derivatives, and `parms` contains the parameter values
(or is `NULL`). The index of the variables is provided in the three-valued vector
`nind`, which contains the number of variables of index 1, 2, and 3 respectively
(see Sect. 4.2.5). Argument `res` is the name of the R function that gets as input the
current `time`, the value of the state variables `y` and their rate of change `dy`, and the
parameter vector, and returns the residuals of the equations.

The syntax for using `radau`, `gamd` or `bimd` is:

```
radau(y, times, func, parms, nind, mass, ...)
gamd (y, times, func, parms, nind, mass, ...)
bimd (y, times, func, parms, nind, mass, ...)
```

where `mass` is the mass matrix M. Note that these functions do not need to be given
initial conditions for the derivatives.

The equations in `res` and `func` must be defined such that the index 1 variables
precede the index 2 variables which in turn precede the index 3 variables.

## 5.2   A Simple DAE of Index 2

Consider the following DAE problem:

$$y'_1 = y_2$$
$$y_1 = \cos(t).$$ (5.3)

The first equation is a simple differential equation. The second equation, which
is an algebraic equation needs two differentiations to obtain an ODE and is thus
of index 2. (first differentiaton gives $y'_1 = y_2 = -\sin(t)$; the second leads to $y'_2 = -\cos(t)$).

The DAE has the analytic solution:

$$y_1 = \cos(t)$$
$$y_2 = -\sin(t).$$ (5.4)

To make the problem well-posed, a consistent set of initial values and initial
derivatives must be supplied:

$$y_1(0) = \cos(0)$$
$$y_2(0) = -\sin(0)$$
$$y_1'(0) = -\sin(0)$$
$$y_2'(0) = -\cos(0).$$

(5.5)

### 5.2.1   Solving the DAEs in General Implicit Form

To solve this problem with mebdfi, we write (5.3) in the general implicit form (see Sect. 4.1):

$$0 = y_1' - y_2$$
$$0 = y_1 - \cos(t),$$

(5.6)

and implement them as:

```
resdae <- function (t, y, dy, p) {
    r1 <- dy[1] - y[2]
    r2 <- y[1] - cos(t)
    list(c(r1, r2))
}
```

Both mebdfi and daspk can solve this higher order DAE written in implicit form. Before invoking this solver, we specify the initial value of the state variables (yini) and their derivatives (dyini), the time sequence (times) and the index of each variable (index) as described in Sect. 4.2.5. The first variable is described by a simple differential equation, and is of index 1; the second variable is of index 2; there is no index 3 variable. We use much looser tolerances than the default (atol, rtol)

```
library(deTestSet)
yini   <- c(y1 = cos(0), y2 = -sin(0))
dyini  <- c(-sin(0), -cos(0))
times  <- seq(from = 0, to = 10, by = 0.1)
index  <- c(1, 1, 0)
out1   <- mebdfi(times = times, res = resdae, y = yini,
                 atol = 1e-10, rtol = 1e-10, dy = dyini,
                 parms = NULL, nind = index)
```

We print the deviation of the numerical solution with the analytic solution:

```
max (abs(out1[,"y1"] - cos(times)),
     abs(out1[,"y2"] + sin(times)))
```

```
[1] 2.349123e-09
```

### *5.2.2   Solving the DAEs in Linearly Implicit Form*

To solve the same problem by `radau`, `gamd` or `bimd`, one must be able to write
the problem in the linearly implicit form $My' = f(t,y,p)$:

$$\begin{pmatrix} 1 & 0 \\ 0 & 0 \end{pmatrix} \cdot \begin{pmatrix} y'_1 \\ y'_2 \end{pmatrix} = \begin{pmatrix} y_2 \\ y_1 - \cos(t) \end{pmatrix}, \tag{5.7}$$

which needs the implementation of the function $f(t,y,p)$ (called `fundae`) and the
mass matrix M:

```
fundae <- function (t, y, p) {
    f1 <- y[2]
    f2 <- y[1] - cos(t)
    list(c(f1, f2))
}
M <- matrix(nrow = 2, ncol = 2, data = c(1, 0, 0, 0))
```

```
out2 <- radau(times = times, fun = fundae, y = yini,
              atol = 1e-10, rtol = 1e-10, mass = M,
              parms = NULL, nind = index)
```

For this problem, `radau` is slightly less accurate than `mebdfi`:

```
max (abs(out2[,"y1"]  - cos(times)),
     abs(out2[,"y2"]  + sin(times)))
```

```
[1] 5.476366e-07
```

The code `gamd` requires the same input parameters as `radau`

```
out3 <- gamd(times = times, fun = fundae, y = yini,
             atol = 1e-10, rtol = 1e-10, mass = M,
             parms = NULL, nind = index)
```

The maximum absolute error for `gamd` is:

```
max (abs(out3[,"y1"]  - cos(times)),
     abs(out3[,"y2"]  + sin(times)))
```

```
[1] 5.90858e-09
```

## 5.3   A Nonlinear Implicit ODE

The next example is a non-linear implicit ODE that can be written in the linearly
implicit form required by `radau`, `gamd` or `bimd` by doubling the number of
equations, obtaining an index one DAE.

However, this nonlinear ODE is solvable, in the original form, by mebdfi and daspk. The equation is:

$$ty^2y'^3 - y^3y'^2 + t(t^2+1)y' - t^2y = 0$$
$$y_{t=1} = \sqrt{3/2},$$

(5.8)

and is solved on the interval t = [1, 10]. This equation has the analytic solution:

$$y = \sqrt{t^2 + 0.5}.$$

(5.9)

The problem is implemented in R as:

```
implicit <- function(t, y, dy, parms) {
    list(t*y^2*dy^3 - y^3*dy^2 + t*(t^2+1)*dy - t^2*y)
}
yini  <- sqrt(3/2)
times <- seq(from = 1, to = 10, by = 0.1)
```

A consistent value for the derivative $y'$ can be found by solving for the root of (5.8) with respect to $y'$, and where $t = 1$ and $y = \sqrt{3/2}$. This is simple enough to be done by hand, but for educational purposes, we use function multiroot from the R package **rootSolve** [11]. We create a function (rootfun) that takes as input the estimate of $y'$ (called dy), and the known values of y and t and returns the value of the equation. As the root will be sought with respect to dy, this should be the first argument in function rootfun; the package **rootSolve** is loaded first:

```
library(rootSolve)
rootfun <- function (dy, y, t)
    t*y^2*dy^3 - y^3*dy^2 + t*(t^2+1)*dy - t^2*y
```

```
dyini <- multiroot(f = rootfun, start = 0, y = yini,
                    t = times[1] )$root
dyini
```

[1] 0.8164966

As this is an ODE, both mebdfi and daspk can solve this problem with the default absolute and relative tolerances (atol=1e-6, rtol=1e-6).

```
out  <- mebdfi(times = times, res = implicit, y = yini,
               dy = dyini, parms = NULL)
out2 <- daspk (times = times, res = implicit, y = yini,
               dy = dyini, parms = NULL)
```

the result obtained by mebdfi is slightly more accurate than the one from daspk.

```
max(abs(out [,2]- sqrt(times^2+0.5)))
```

[1] 3.017694e-06

```
max(abs(out2[,2]- sqrt(times^2+0.5)))
```

```
[1] 5.689474e-05
```

To solve the same problem with gamd, bimd or radau we rewrite (5.8) in linearly implicit form by adding the variable $z = y'$. The problem implemented in R is:

```
implicit2 <- function (t, y, p) {
    f1 <- y[2]
    f2 <- t*y[1]^2*y[2]^3-y[1]^3*y[2]^2+t*(t^2+1)*y[2]-t^2*y[1]
    list(c(f1, f2))
}
M <- matrix(nrow = 2, ncol = 2, data = c(1, 0, 0, 0))
yini_li       <- c(yini, dyini)
```

Solving it with bimd

```
out3 <- bimd(times = times, fun = implicit2, y = yini_li,
             mass = M, parms = NULL)
```

we obtain the following absolute error:

```
max(abs(out3[,2]- sqrt(times^2+0.5)))
```

```
[1] 5.709242e-07
```

We note that the size of the problems is now doubled with respect to the original nonlinear implicit ODE.

## 5.4   A DAE of Index 3: The Pendulum Problem

We now implement the pendulum problem of (Chap. 4, Sect. 4.1.5):

$$\begin{aligned}
x' &= u \\
y' &= v \\
u' &= -\lambda x \\
v' &= -\lambda y - g \\
0 &= x^2 + y^2 - L^2.
\end{aligned} \tag{5.10}$$

In the R implementations of (5.10), we take $L = 1$ and $g = 9.8$. We implement the index 3 system only, and leave the index 1 (4.41) and 2 (4.40) DAE models to be implemented by the reader.

```
library(deTestSet)
pendulum <- function (t, y, dy, parms) {
   list(c(-dy[1] + y[3]           ,
```

```
              -dy[2] + y[4]              ,
              -dy[3] -y[5]*y[1]          ,
              -dy[4] -y[5]*y[2]  - 9.8,
               y[1]^2 + y[2]^2 -1
       ))
 }
```

where $x, y, u, v$ and $\lambda$ are y[1], y[2] ...y[5] respectively.

When solving this equation it is clear that the initial values $(x_0, y_0)$ have to satisfy the constraint $x_0^2 + y_0^2 = L^2$. We assume that the pendulum is swung horizontally to the right. The initial conditions for the derivatives $x'$, $y'$, $u'$, and $v'$ can be derived from the initial values, based on the differential equations. The initial condition for $\lambda$ can be estimated from (4.41) (Sect. 4.2.4.1). The initial conditions for the derivative of $\lambda$ can only be estimated after the algebraic equation has been differentiated three times! (see Sect. 4.2.4.1, equation (4.41)). The first two variables are of index 1, followed by two  variables of index 2, then there is one index 3  variable. This information is incorporated in argument index3.

```
yini  <- c(x = 1,  y = 0, u = 0,  v = 1  , lam = 1)
dyini <- c(dx = 0,dy = 1,du = -1,dv = -9.8,dlam = 3*9.8)
times <- seq(from = 0, to = 10, by = 0.01)
index3 <- c(2,  2,  1)
out3 <- mebdfi (y = yini, dy = dyini, res = pendulum,
                parms = NULL, times = times,
                nind = index3)
```

The output (Fig. 5.1) shows the periodic behavior of all variables, and the swing of the ball (last figure).

```
plot(out3, lwd = 2)
plot(out3[, 2:3])
mtext(side = 3, outer = TRUE, line = -1.5,
      "Pendulum", cex = 1.5)
```

## 5.5   Multibody Systems

A multibody system is a mechanical system consisting of interconnected rigid or elastic bodies. They can move relative to one another, but are connected. The connections can be either force elements such as springs and dampers, or friction or joints. The latter restrict relative movement of pairs of bodies [15].

The equations describing multibody systems are the equation of motion, (Newton's 2nd law), amended with a restriction, caused by the different joints:

$$M(q)q'' = f(q, q')$$
$$g(q) \quad = 0, \tag{5.11}$$

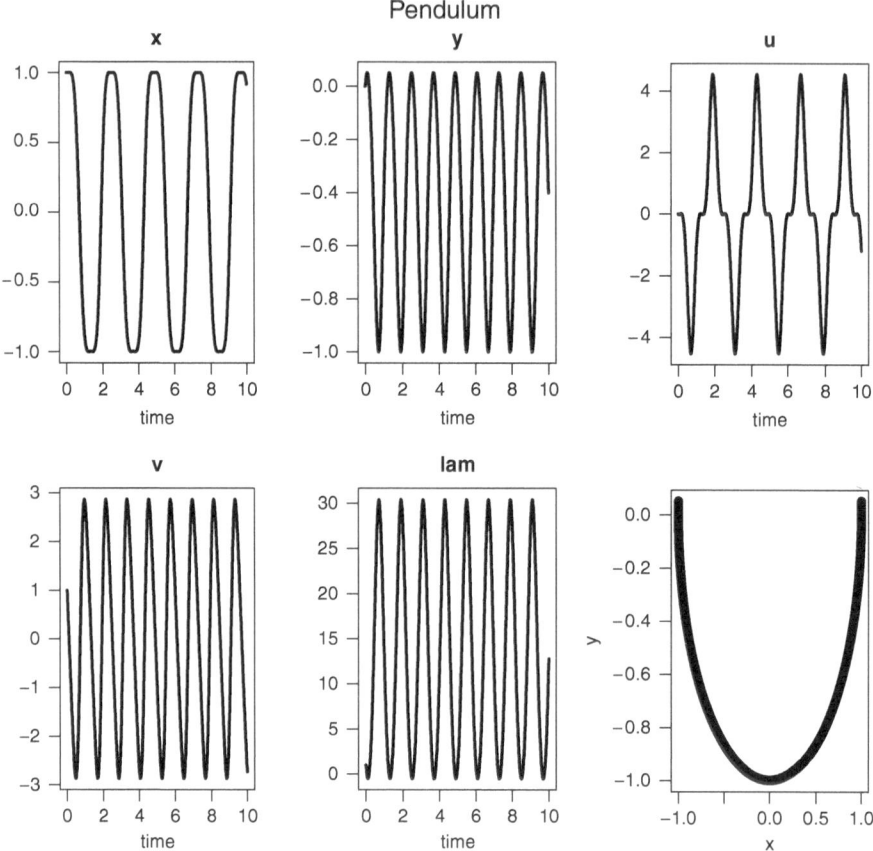

**Fig. 5.1** Solution of the pendulum problem, See text for the R code

where $M$ is the mass matrix, $q$ denotes the position of the bodies, and $f$ are the internal and external forces acting on the system. Forces can be a function of both the position ($q$) of the bodies, e.g. elastic forces, or their velocity ($q'$), e.g. frictional forces. The restriction, $g$, only acts on the position.

The pendulum problem from Sect. 4.1.5 was a rather simple example of a multibody system.

## 5.5.1  The Car Axis Problem

We now implement the car axis problem from the test set for initial value problem solvers [8] (Fig. 5.2). This is a stiff DAE, of index 3, consisting of eight differential and two algebraic equations. It models the behavior of a car axis on a bumpy road. The first four variables represent the positions of the left and right springs, attached

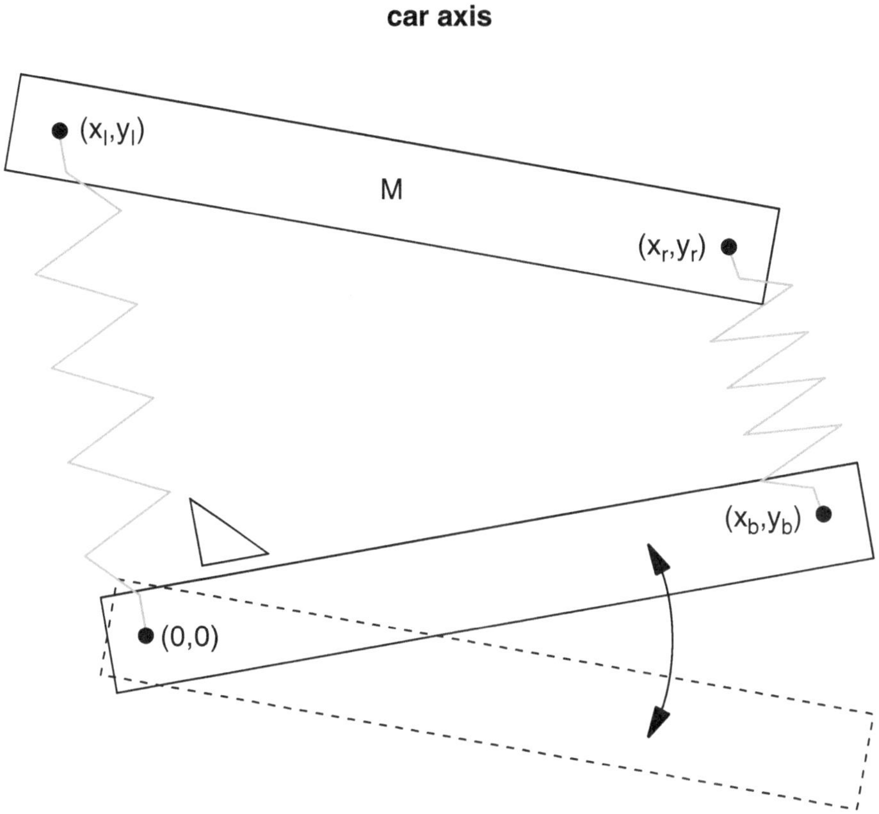

**car axis**

**Fig. 5.2** Schematic respresentation of the car axis problem

to the wheels, $(x_l, y_l, x_r, y_r)$, the next four variables represent the velocity vector $(u_l, v_l, u_r, v_r)$, and the last two variables $(\lambda_1, \lambda_2)$ relate to the two position constraints.

The parameter settings are such that the problem describes the situation where the left wheel rolls on a flat surface, while the right wheel rolls over a bumpy surface, whose height is described by the function $r \sin wt$. The coordinates of the left wheel are assumed to be $(0,0)$. The length of the car axis (distance between two wheels) is $L$, and remains fixed, such that at all times, the coordinates of the right wheel are described by $(x_b, y_b)$ where $y_b = r \sin wt$ and $x_b = \sqrt{L^2 - y_b^2}$.

The wheels are connected to the chassis of the car with mass $M$, by means of two springs, attached at positions $(x_l, y_l)$ and $(x_r, y_r)$ for the left and right side respectively. It is assumed that the springs are of zero mass, and have length $L_0$ at rest. Two constraints restrict the independent movement of the bodies. The first constraint specifies that the left spring remains orthogonal to the axis

$$x_l x_b + y_l y_b = 0. \tag{5.12}$$

The second constraint imposes the condition that the distance between the two spring attachments remain at constant length $L$.

$$(x_l - x_r)^2 + (y_l - y_r)^2 = L^2. \tag{5.13}$$

The second order equations for the positions of the spring $x_l, y_l, x_r, y_r$ are given by:

$$\begin{aligned}
x_l'' &= 1/k[(L_0 - L_l)\frac{x_l - 0}{L_l} + 2\lambda_2(x_l - x_r) + \lambda_1(x_b - 0)] \\
y_l'' &= 1/k[(L_0 - L_l)\frac{y_l - 0}{L_l} + 2\lambda_2(y_l - y_r) + \lambda_1(y_b - 0)] - g \\
x_r'' &= 1/k[(L_0 - L_r)\frac{x_r - x_b}{L_r} - 2\lambda_2(x_l - x_r)] \\
y_r'' &= 1/k[(L_0 - L_r)\frac{y_r - y_b}{L_r} - 2\lambda_2(y_l - y_r)] - g,
\end{aligned} \tag{5.14}$$

where $\lambda_1$ and $\lambda_2$ are the constraint forces (they are also Lagrange multipliers) and $L_l$ and $L_r$ are the lengths of the left and right springs respectively.

Equations (5.14) are rewritten as a system of first order differential equations (not shown) by also modelling the velocity vector $(u_l, v_l, u_r, v_r)$, where $u$ and $v$ are the velocity in the $x$ and $y$ directions respectively. Added to these eight first order differential equations are the two constraints (5.12) and (5.13).

We implement these equations in an R function, caraxis, that takes as input the time (t), the values of the variables (y) and their derivatives (dy), and the parameters (parms, not used).

The function returns the residuals (delt) packed as a list.

To make the model readable, we use the *names* of the dependent variables; with (as.list(y), ...) does that.

```
caraxis <- function(t, y, dy, parms) {
  with(as.list(y), {
    f <- rep(0, 10)
    yb <- r * sin(w * t)
    xb <- sqrt(L^2  - yb^2)
    Ll <- sqrt(xl^2 + yl^2)
    Lr <- sqrt((xr - xb)^2 + (yr - yb)^2)
    f[1:4]  <- y[5:8]
    f[5]  <- 1/k*((L0-Ll)*xl/Ll  + lam1*xb  + 2*lam2*(xl-xr))
    f[6]  <- 1/k*((L0-Ll)*yl/Ll  + lam1*yb  + 2*lam2*(yl-yr))  -g
    f[7]  <- 1/k*((L0-Lr)*(xr  - xb)/Lr   - 2*lam2*(xl-xr))
    f[8]  <- 1/k*((L0-Lr)*(yr  - yb)/Lr   - 2*lam2*(yl-yr))  -g
    f[9]  <- xb * xl + yb * yl
    f[10]<- (xl  - xr)^2 + (yl  - yr)^2  - L^2

    delt       <- dy - f
    delt[9:10] <- -f[9:10]

    list(delt)
  })
}
```

The parameter values are:

```
eps <- 0.01; M <- 10; k <- M * eps * eps/2
L <- 1; L0 <- 0.5; r <- 0.1; w <- 10; g <- 9.8
```

and the initial conditions

```
yini <- c(xl = 0,      yl = L0, xr = L,     yr = L0,
          ul = -L0/L, vl = 0,   ur = -L0/L, vr = 0,
          lam1 = 0, lam2 = 0)
```

Similarly to what we did in the previous example, we use a root solving procedure to retrieve a consistent set of initial derivatives (dyini) for this problem. From the equations, it is clear that the derivatives of the last two state variables $\lambda_1, \lambda_2$, (lam1, and lam2) do not affect the residuals. Hence, they are kept out of the root-finding procedure, which solves for the first eight variables only. We arbitrarily set these derivatives equal to 0.

```
library(rootSolve)
rootfun <- function (dyi, y, t)
    unlist(caraxis(t, y, dy = c(dyi, 0, 0), parms = NULL)) [1:8]
```

```
dyini <- multiroot(f = rootfun, start = rep(0,8),
                   y = yini, t = 0)$root
(dyini <- c(dyini,0,0))
```

```
[1] -0.500000  0.000000 -0.500000  0.000000  0.000000
[6] -9.799999  0.000000 -9.799999  0.000000  0.000000
```

We check the consistency of the initial values, to be sure:

```
caraxis(t = 0, yini, dyini, NULL)
```

```
[[1]]
[1] 2.512380e-09 0.000000e+00 2.512380e-09 0.000000e+00
[5] 0.000000e+00 8.108556e-07 0.000000e+00 8.108556e-07
[9] 0.000000e+00 0.000000e+00
```

Next the index of the system variables is defined, and the problem solved, using mebdfi:

```
index <- c(4, 4, 2)
times <- seq(from = 0, to = 3, by = 0.01)
out <- mebdfi(y = yini, dy = dyini, times = times,
              res = caraxis, parms = parameter, nind = index)
```

The results (Fig. 5.3) first depict the positions and velocities versus time and then show how the various springs move. First the margins of each figure are reduced (mar).

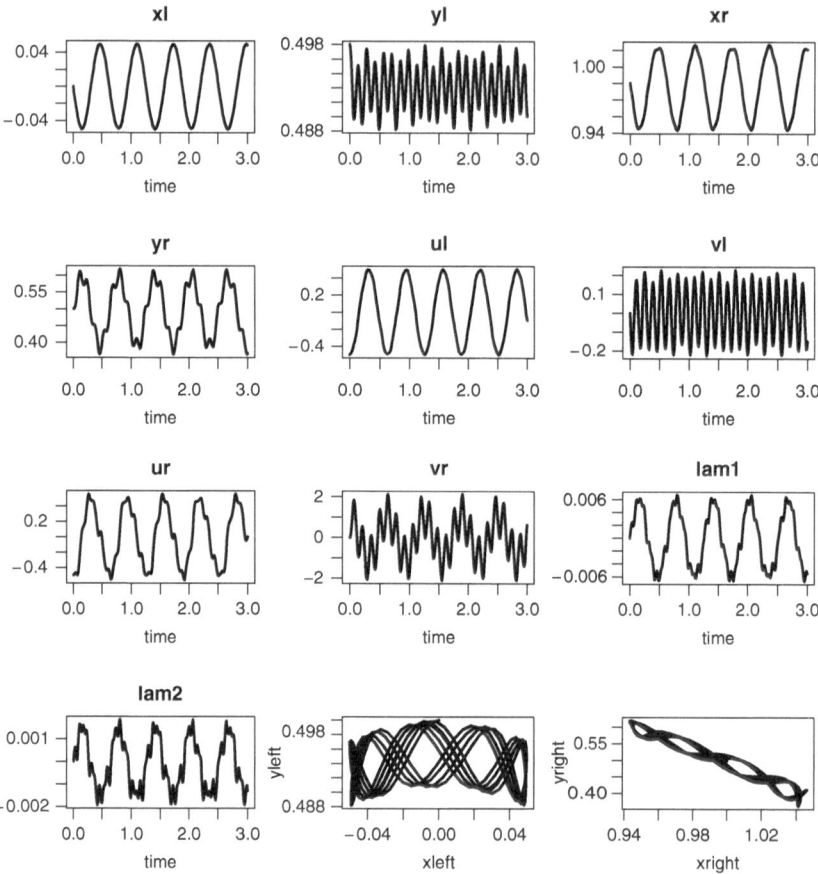

**Fig. 5.3** The car axis problem solution. See text for the R code

```
par(mar = c(4, 4, 3, 2))
plot(out, lwd = 2, mfrow = c(4,3))
plot(out[,c("xl", "yl")], xlab = "xleft", ylab = "yleft",
    type = "l", lwd = 2)
plot(out[,c("xr", "yr")], xlab = "xright", ylab = "yright",
    type = "l", lwd = 2)
```

## 5.6  Electrical Circuit Models

Mathematical models are often used to assess the behavior of electrical circuits before actually producing electronic devices [5,14]. These models combine physical laws such as energy and charge conservation with the characteristics of the network elements. Often the resulting model consists of a differential algebraic system.

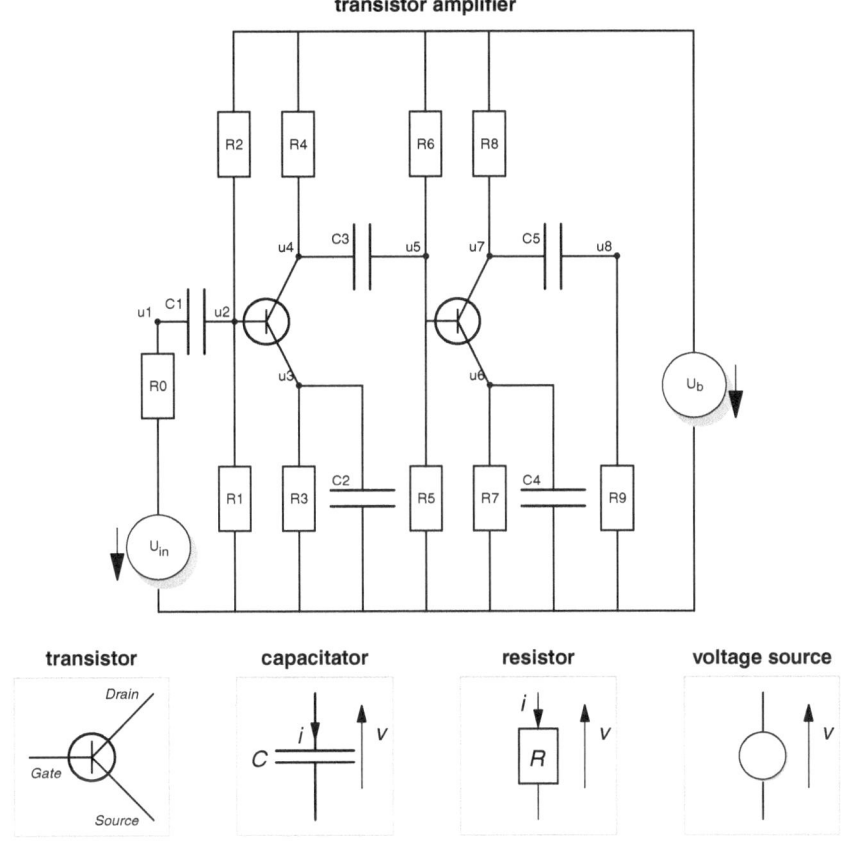

**Fig. 5.4** The transistor amplifier

An electrical network model consists of elements and nodes. The composition of these elements obeys Kirchoff's voltage and current law, which (simplified) says:

- For each loop of the network, the algebraic sum of voltages is zero.
- For each node, the sum of all currents is zero.

Kirchoff's voltage law can be used to obtain a relationship between branch voltages, $v(t)$, and node voltages, $u(t)$, which for a branch connecting node $j$ and $j+1$ gives:

$$v(t) = u(t)_j - u(t)_{j+1}. \tag{5.15}$$

### 5.6.1 The Transistor Amplifier

The transistor amplifier circuit [10] of Fig. 5.4 is a frequently used benchmark circuit for testing DAE solvers [8].

The circuit consists of eight nodes. A voltage source produces an input signal given by $(U_{in} = 0.1\sin(200\pi t))$, and which is amplified by two transistors. The node potential at node 8 consists of the output signal.

Each transistor is modeled as:

$$
\begin{aligned}
I_G &= (1 - \alpha)g(u_g - u_s) \\
I_D &= \alpha g(u_g - u_s) \\
I_S &= g(u_g - u_s) \\
g(u_g - u_s) &= \beta\left(\exp\left(\frac{u_g - u_s}{u_f}\right) - 1\right),
\end{aligned}
\tag{5.16}
$$

where $I_G$, $I_D$ and $I_S$ are the currents through the gate, drain and source contact, respectively. The transistor parameters are: $\beta = 10^{-6}$, $u_f = 0.026$, $\alpha = 0.99$.

In addition, there are nine resistors, and five capacitators in the network, obeying Ohm's and Faraday's law respectively:

$$
I_R = \frac{V_R}{R}
\tag{5.17}
$$

$$
I_C = C\frac{dV_c}{dt} = CV_c'.
\tag{5.18}
$$

The circuit model calculates the transient (time-dependent) behavior of the electrical signal at each node, in response to the time-varying input signal. If we denote with $I$, $V$ the branch currents and voltage drops and with $u$ the node potential, then applying Kirchoff's current law for the second node gives:

$$
0 = I_{C_1} - I_{R_1} - I_{R_2} - I_{gate,2}.
\tag{5.19}
$$

Using (5.16)–(5.18) this can be expanded as:

$$
\begin{aligned}
0 &= C_1 V_{C_1}' - \frac{V_{R_1}}{R_1} - \frac{V_{R_2}}{R_2} - (1 - \alpha)g(u_2 - u_3) \\
0 &= C_1(u_2 - u_1)' - \frac{u_2}{R_1} - \frac{(u_2 - U_b)}{R_2} - (1 - \alpha)g(u_2 - u_3).
\end{aligned}
\tag{5.20}
$$

$U_b$ is the working voltage of the circuit, which is set to 6. The other parameters are chosen to be $R_0 = 1,000$; $R_i = 9,000$ for $i = 1,\ldots,9$ and $C_1 = 10^{-6}$, $C_2 = 2\cdot 10^{-6}$, $\ldots$ $C_6 = 6\cdot 10^{-6}$.

The equations for the entire network can be compactly represented as:

$$
Mu' = f(t,u),
\tag{5.21}
$$

where matrix $M$ is given by:

$$
\begin{pmatrix}
-C_1 & C_1 & 0 & 0 & 0 & 0 & 0 & 0 \\
C_1 & -C_1 & 0 & 0 & 0 & 0 & 0 & 0 \\
0 & 0 & -C_2 & 0 & 0 & 0 & 0 & 0 \\
0 & 0 & 0 & -C_3 & C_3 & 0 & 0 & 0 \\
0 & 0 & 0 & C_3 & -C_3 & 0 & 0 & 0 \\
0 & 0 & 0 & 0 & 0 & -C_4 & 0 & 0 \\
0 & 0 & 0 & 0 & 0 & 0 & -C_5 & C_5 \\
0 & 0 & 0 & 0 & 0 & 0 & C_5 & -C_5
\end{pmatrix},
\tag{5.22}
$$

and function $f$ is:

$$
\begin{pmatrix}
\dfrac{u_1 - U_e(t)}{R_0} \\[2ex]
\dfrac{u_2}{R_1} + \dfrac{u_2 - U_b}{R_2} + (1-\alpha)g(u_2 - u_3) \\[2ex]
\dfrac{u_3}{R_3} - g(u_2 - u_3) \\[2ex]
\dfrac{u_4 - U_b}{R_4} + \alpha g(u_2 - u_3) \\[2ex]
\dfrac{u_5}{R_5} + \dfrac{u_5 - U_b}{R_6} + (1-\alpha)g(u_5 - u_6) \\[2ex]
\dfrac{u_6}{R_7} - g(u_5 - u_6) \\[2ex]
\dfrac{u_7 - U_b}{R_8} + \alpha g(u_5 - u_6) \\[2ex]
\dfrac{u_8}{R_9}
\end{pmatrix}.
\tag{5.23}
$$

The derivation of the complete model equations can be found in [8].

Here it is implemented in linearly implicit form and solved with radau:

```
library(deSolve)
Transistor <- function(t, u, du, pars) {
    delt <- vector(length = 8)
    uin  <- 0.1 * sin(200 * pi * t)
    g23  <- beta * (exp( (u[2]  - u[3]) / uf) - 1)
    g56  <- beta * (exp( (u[5]  - u[6]) / uf) - 1)

    delt[1] <- (u[1]  - uin)/R0
    delt[2] <- u[2]/R1 + (u[2]-ub)/R2 + (1-alpha) * g23
    delt[3] <- u[3]/R3 - g23
    delt[4] <- (u[4]  - ub) / R4 + alpha * g23
    delt[5] <- u[5]/R5 + (u[5]-ub)/R6 + (1-alpha) * g56
    delt[6] <- u[6]/R7 - g56
    delt[7] <- (u[7]  - ub) / R8 + alpha * g56
    delt[8] <- u[8]/R9
    list(delt)
}
```

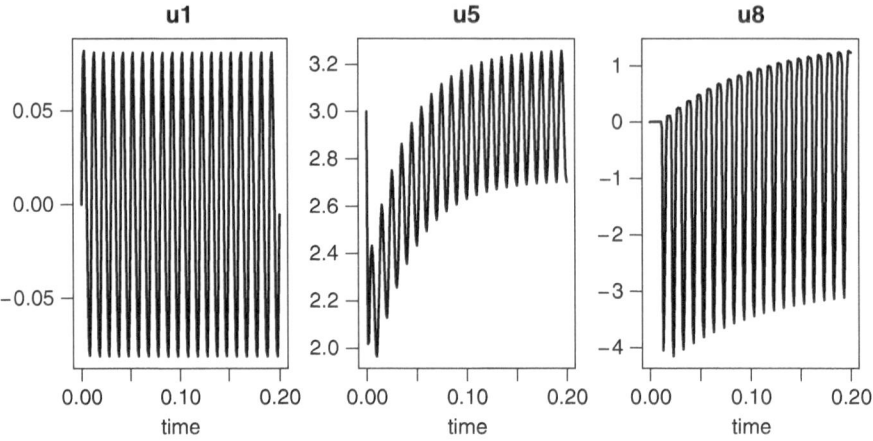

**Fig. 5.5** The transistor problem solution; potentials at node 1, 5, and 8. See text for the R code

```
ub <- 6; uf <- 0.026; alpha <- 0.99; beta <- 1e-6; R0 <- 1000
R1 <- R2 <- R3 <- R4 <- R5 <- R6 <- R7 <- R8 <- R9 <- 9000
C1 <- 1e-6; C2 <- 2e-6; C3 <- 3e-6; C4 <- 4e-6; C5 <- 5e-6
```

```
mass <- matrix(nrow = 8, ncol = 8, byrow = TRUE, data = c(
      -C1,C1, 0,  0,  0,  0,  0,  0,
      C1,-C1, 0,  0,  0,  0,  0,  0,
      0,  0,-C2,  0,  0,  0,  0,  0,
      0,  0,  0,-C3, C3,  0,  0,  0,
      0,  0,  0, C3,-C3,  0,  0,  0,
      0,  0,  0,  0,  0,-C4,  0,  0,
      0,  0,  0,  0,  0,  0,-C5, C5,
      0,  0,  0,  0,  0,  0, C5,-C5
))
```

```
yini <- c(0, ub/(R2/R1+1), ub/(R2/R1+1),
          ub, ub/(R6/R5+1), ub/(R6/R5+1), ub, 0)
names(yini) <- paste("u", 1:8, sep = "")
```

```
ind   <- c(8, 0, 0)
times <- seq(from = 0, to = 0.2, by = 0.001)
out <- radau(func = Transistor, y = yini, parms = NULL,
             times = times, mass = mass, nind = ind)
```

```
plot(out, lwd = 2, which = c("u1", "u5", "u8"),
     mfrow = c(1, 3))
```

## 5.7 Exercises

### 5.7.1 A Simple DAE

You are given the following simple DAE:

$$\begin{aligned}
0 &= 2y_1 + 3y_2 \\
0 &= 3y_1 - y_3 - 4 \\
y_3' &= y_1 + y_2,
\end{aligned} \tag{5.24}$$

with initial conditions $y_1(0) = 3, y_1'(0) = 0, y_2 = -2, y_2' = 0, y_3 = 5, y_3' = 1$. Implement this problem, and solve it with the five DAE solvers for $t \in [0,10]$. The values at $t = 10$ are $9.113195, -6.075464$ and $23.33959$ for $y_1, y_2, y_3$ respectively.

### 5.7.2 The Robertson Problem

The Robertson problem is a classic problem to test stiff ODE solvers. You were asked to solve the system of ODEs in Sect. 3.6.2. Here, you are asked to solve the problem written as a DAE:

$$\begin{aligned}
y_1' &= -0.04y_1 + 10^4 y_2 y_3 \\
y_2' &= 0.04y_1 - 10^4 y_2 y_3 - 3 \cdot 10^7 y_2^2 \\
1 &= y_1 + y_2 + y_3.
\end{aligned} \tag{5.25}$$

The initial conditions are $(y_1 = 1, y_2 = 0, y_3 = 0)$. Implement this DAE; solve it with radau and with daspk. When using daspk, implement it in fully implicit form. Integrate the problem on the interval $0 \le t \le 40$. As it is much more challenging to solve this equation on a much larger interval (e.g. [6, p. 144], you should also try to integrate it in the interval $10^{-4} \le t \le 10^7$. Use for the second output times a logarithmic series, as shown in Sect. 3.6.2.

### 5.7.3 The Pendulum Problem Revisited

In Sect. 5.4 we implemented the index 3 pendulum problem in implicit form and solved it with mebdfi. Rewrite this problem in linearly implicit form and solve it with radau.

Implement the different problem formulations of index 1 and 2 (see Sect. 4.2.4.1). The error in the DAE solution can be assessed by computing how much the algebraic condition is violated. Create a figure that represents this error for the different implementations (see Fig. 4.2)

### 5.7.4  The Akzo Nobel Problem

This last exercise is considerably more complex than the previous ones. The example originates from the Akzo Nobel Central Research in Arnhem, the Netherlands and is one of the test problems in [8]. The problem formulation is slightly modified so as to show how to derive the DAEs that arise from equilibrium chemistry.[2]

#### 5.7.4.1  Problem Formulation

Two chemical species $B$ and $Z$ are mixed in a vessel, while carbon dioxide $CO_2$ is continuously added. In a series of reactions, $A$ is produced and this is the species of importance. The names of the species, except for nitrate, $H_2O$ and $CO_2$ are fictituous. The reactions are:

$$
\begin{aligned}
2B + 0.5CO_2 &\xrightarrow{r_1} BT + H_2O \\
A + B &\underset{r_3}{\overset{r_2}{\rightleftharpoons}} BT + Z \\
B + 2Z + CO_2 &\xrightarrow{r_4} LB + nitrate \\
B.Z + 0.5CO_2 &\xrightarrow{r_5} A + H_2O \\
B + Z &\underset{r_7}{\overset{r_6}{\rightleftharpoons}} B.Z,
\end{aligned}
\tag{5.26}
$$

where $r_1, \ldots, r_7$ are the reaction rates (units of concentration per time). Based on these chemical processes, we can write a mass balance for the concentration of each species:

$$
\begin{aligned}
[B]' &= -2r_1 - r_2 + r_3 - r_4 - r_6 + r_7 \\
[CO2]' &= -0.5r_1 - r_4 - 0.5r_5 + E \\
[BT]' &= r_1 + r_2 - r_3 \\
[Z]' &= r_2 - r_3 - 2r_4 - r_6 + r_7 \\
[A]' &= -r_2 + r_3 + r_5 \\
[B.Z]' &= -r_5 + r_6 - r_7,
\end{aligned}
\tag{5.27}
$$

where $[X]$ denotes the concentration of species $X$, and $E$ is the exchange rate of $CO_2$ with the atmosphere. According to the mass action law [1], the speed of a chemical reaction is proportional to the product of the concentrations of the participating molecules where we take into account the reaction's stochiometric coefficients. For instance, for the first reaction, with rate $r_1$, two moles of $B$ react with 0.5 moles of $CO_2$ to produce $BT$ and water. Thus the reaction will be proportional to

---

[2]And to distinguish it from the test problem in [8] which is different.

$[B]^2[CO_2]^{1/2}$. Based on these principles, all reaction velocities are given by[3]:

$$r_1 = k_1[B]^2[CO_2]^{1/2}$$
$$r_2 = k_2[A][B]$$
$$r_3 = k_3[BT][Z]$$
$$r_4 = k_4[B][Z]^2[CO_2] \qquad (5.28)$$
$$r_5 = k_5[B.Z][CO_2]^{1/2}$$
$$r_6 = k_6[B][Z]$$
$$r_7 = k_7[B.Z],$$

where $k_i$ are the rate constants. The exchange of $CO_2$ with the overlying atmosphere is described as:

$$E = k_A(pCO_2/k_H - [CO_2]), \qquad (5.29)$$

with $k_A$ the mass transfer coefficient, $k_H$ Henry's constant [16], and $pCO_2$ the partial pressure of $CO_2$.

The combination of (5.27)–(5.29) define the problem as a set of ODEs. Before this system of equations can be solved, values for all rate constants ($k_i$, $k_A$, …) are needed. In general, it is feasible to measure the rate constants for reactions that are sufficiently slow, but it is much more difficult to do so for very fast reactions.

In the Akzo Nobel problem, the reaction velocities $r_6$ and $r_7$ are so high that the reactions can be considered to operate at local equilibrium. That is, the *local* rate of change of Z, B or B.Z due to this reaction is assumed to be equal to 0. For the concentration of Z this gives:

$$\frac{d[Z]}{dt}\bigg|_{eq} = 0 = -r_6 + r_7. \qquad (5.30)$$

Combining $r_6 = r_7$ with (5.28) leads to

$$k_6[B][Z] = k_7[B.Z]. \qquad (5.31)$$

If we define $K_s = k_6/k_7$, this gives way to the algebraic equation:

$$K_s = \frac{[B.Z]}{[B][Z]}, \qquad (5.32)$$

where $K_s$ is the equilibrium constant. In contrast to $k_6$ and $k_7$ whose values are very difficult to determine, it is relatively simple to measure $K_s$.

---

[3]This differs from the original description.

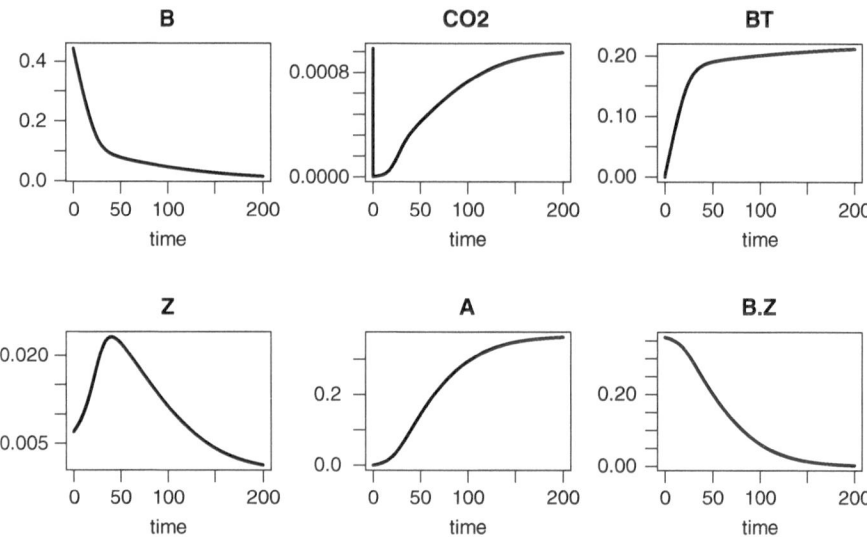

**Fig. 5.6** The Akzo nobel problem

As the equilibrium formulation (5.32) has removed the constants $k_6$ and $k_7$, the corresponding rates $r_6$ and $r_7$ should now also be removed from the mass balances (5.27). This is done by taking suitable linear combinations of (5.27). There are several ways to achieve this. The most logical is to add the mass balance of $[B]$ and $[B.Z]$ and of $[Z]$ and $[B.Z]$ (in 5.27) to obtain:

$$[B]' + [B.Z]' = -2r_1 - r_2 + r_3 - r_4 - r_5$$
$$[CO2]' = -0.5r_1 - r_4 - 0.5r_5 + E$$
$$[BT]' = r_1 + r_2 - r_3 \qquad (5.33)$$
$$[Z]' + [B.Z]' = r_2 - r_3 - 2r_4 - r_5$$
$$[A]' = -r_2 + r_3 + r_5.$$

#### 5.7.4.2  Task

The set of (5.33) with the constraint (5.32) forms the DAE system to be solved. Your task is to implement the model, and solve it with `radau`, using the following initial conditions: $[B] = 0.444 \ \text{mol} \ \text{l}^{-1}, [Z] = 0.007 \ \text{mol} \ \text{l}^{-1}, [CO_2] = 0.00123 \ \text{mol} \ \text{l}^{-1}$, other species are initially not present.

The various constants have the value: $k_1 = 18.7; \ k_2 = 0.58/34.4; \ k_3 = 0.58; \ k_4 = 0.09; \ k_5 = 0.42; \ K_s = 115.83; \ p(CO2) = 0.9; \ k_A = 3.3; \ k_H = 737$

The results of this model are in Fig. 5.6:

# References

1. Aris, R. (1965). *Introduction to the analysis of chemical reactors*. Englewood Cliffs: Prentice Hall.
2. Brenan, K. E., Campbell, S. L., & Petzold, L. R. (1996). *Numerical Solution of Initial-Value Problems in Differential-Algebraic Equations. SIAM classics in applied mathematics*. Philadelphia, PA: SIAM.
3. Brugnano, L., Magherini, C., & Mugnai, F. (2006). Blended implicit methods for the numerical solution of DAE problems. *Journal of Computational and Applied Mathematics, 189*(1–2), 34–50.
4. Cash, J. R., & Considine, S. (1992). An **MEBDF** code for stiff initial value problems. *ACM Transactions on Mathematical Software, 18*(2), 142–158.
5. Günther, M., Feldmann, U., & ter Maten, E. J. W. (2005). Modelling and discretization of circuit problems. In W. H. A. Schilders & E. J. W. ter Maten (Eds.), *Numerical analysis in electromagnetics* (pp. 523–659). Amsterdam: North-Holland/Elsevier.
6. Hairer, E., & Wanner, G. (1996). *Solving ordinary differential equations II: Stiff and differential-algebraic problems*. Heidelberg: Springer.
7. Iavernaro, F., & Mazzia, F. (1998). Solving ordinary differential equations by generalized Adams methods: properties and implementation techniques. *Applied Numerical Mathematics, 28*(2–4), 107–126. Eighth Conference on the Numerical Treatment of Differential Equations (Alexisbad, 1997).
8. Mazzia, F., & Magherini, C. (2008). *Test set for initial value problem solvers, release 2.4* (Rep. 4/2008). Department of Mathematics, University of Bari, Italy.
9. Petzold L. R. (1983). A description of **DASSL**: A differential/algebraic system solver. *IMACS Transactions on Scientific Computation*, New Brunswick, NJ, pp. 65–68.
10. Rentrop, P., Roche, M., & Steinebach, G. (1989). The application of Rosenbrock-Wanner type methods with stepsize control in differential-algebraic equations. *Numerische Mathematik, 55*, 545–563.
11. Soetaert, K. (2011). **rootSolve***: Nonlinear root finding, equilibrium and steady-state analysis of ordinary differential equations*. R package version 1.6.2.
12. Soetaert, K., Petzoldt, T., & Setzer, R. W. (2010). Solving differential equations in R: Package **deSolve**. *Journal of Statistical Software, 33*(9), 1–25.
13. Soetaert, K., Cash, J. R., & Mazzia, F. (2011). **deTestSet***: Testset for differential equations*. R package version 1.0.
14. Voigtmann, S. (2006). *General linear methods for integrated circuit design*. PhD thesis, Humboldt-Universitat zu Berlin. (Online: Stand 2010-07-03T11:52:37Z).
15. Wagner, F. J. (1999). Multibody systems. In C. Bendtsen & P. G. Thomsen (Eds.), *Numerical solution of differential algebraic equations* (Tech. Rep. IMM-REP-1999-8). Lyngby, Denmark: IMM, Department of Mathematical Modelling, Technical University of Denmark. Chapter 9.
16. Zeebe, R. E., & Wolf-Gladrow, D. (2001). *$CO_2$ in Seawater: Equilibrium, kinetics, isotopes. Elsevier oceanography series*. Amsterdam: Elsevier.

# Chapter 6
# Delay Differential Equations

**Abstract** Delay differential equations (DDEs) are similar to ordinary differential equations, except that they involve *past* values of the dependent variables and/or their derivatives. Because of this, rather than needing an initial *value* to be fully specified, DDEs require input of an initial *history* (sequence of values) instead. Typically, these initial (history) functions are not fully compatible with the model dynamics, leading to discontinuities as the method switches from the initial function to values recorded through the integration. Whether the delay is introduced in the values or in the derivatives has great implications for the propagation in time of the discontinuities.

## 6.1 Delay Differential Equations

Delay differential equations arise in many different areas of science. If an action is to be made based on an assessment of the current state of a system and if some time is necessary to process the information, the action will not be taken instantaneously but rather a delay will arise. This delay is best incorporated in differential equations by making the action a function of past rather than of instantaneous values of the dependent variables.

A survey of the use of DDEs [1] indicates that most arise from biological models although they are applied throughout the sciences. In natural sciences, delay differential equations have been used for example in the modelling of El Niño temperature oscillations in the Equatorial Pacific [13], or to model single-species population growth [9]. In electrical circuits, delays are introduced because it takes time for a signal to travel through a transmission line.

From a mathematical point of view, delay differential equations (DDEs) differ from ordinary differential equations because the evolution of DDEs involves a time series of *past* values of dependent variables and derivatives, whereas the evolution of ODEs depends only on the *current* values of these quantities. For more information about the background theory related to existence and regularity of solutions of

K. Soetaert et al., *Solving Differential Equations in R*, Use R!,
DOI 10.1007/978-3-642-28070-2_6, © Springer-Verlag Berlin Heidelberg 2012

DDEs and to their numerical integration, we refer the interested reader to the comprehensive books [3, 8] and to the wide bibliography therein included, as well as to the more recent and updated survey paper [4] devoted to delay differential equations and more general functional differential equations.

### 6.1.1  DDEs with Delays of the Dependent Variables

When depending on past *values* of the dependent variables only, DDEs are sometimes called *retarded* delay differential equations (RDDE). They can be written as:

$$y'(t) = f(t, y(t), y(t - \tau_1), y(t - \tau_2), \ldots, y(t - \tau_n)), \qquad \text{for} \quad t_0 \le t \le t_F$$
$$y(t) = \Phi(t), \qquad \text{for} \quad t \le t_0, \tag{6.1}$$

where $y'$ is the (right-hand) derivative of $y$ with respect to $t$, $\tau$ is called the delay, $(t - \tau)$ the delay argument and $y(t - \tau)$ the delay value. The function $\Phi(t)$ provides the history of the dependent variables before the start of the simulation.

This formalism assumes fixed delays. A more general representation is:

$$y'(t) = f(t, y(t), y(\alpha_1(t, y(t))), \ldots, y(\alpha_n(t, y(t)))), \qquad \text{for} \quad t_0 \le t \le t_F$$
$$y(t) = \Phi(t), \qquad \text{for} \quad t \le t_0, \tag{6.2}$$

where the delay functions $\alpha_i(t, y(t))$, $i = 1, \ldots, n$ satisfy $\alpha_i(t, y(t)) \le t$.

### 6.1.2  DDEs with Delays of the Derivatives

DDEs where the differential equations also depend on past *derivatives* of the dependent variables are called *neutral* delay differential equations (NDDE). They are given by:

$$y'(t) = f(t, y(t), y(t - \tau_1), \ldots, y(t - \tau_n), y'(t - \sigma_1), \ldots, y'(t - \sigma_m)),$$
$$\text{for} \quad t_0 \le t \le t_F$$
$$y(t) = \Phi(t), \qquad \text{for} \quad t \le t_0$$
$$y'(t) = \Phi'(t), \qquad \text{for} \quad t \le t_0, \tag{6.3}$$

for fixed delays, or more generally:

$$y'(t) = f(t, y(t), y(\alpha_1(t, y(t))), \ldots, y(\alpha_n(t, y(t))), y'(\beta_1(t, y(t))), \ldots, y'(\beta_m(t, y(t)))),$$
$$\text{for} \quad t_0 \le t \le t_F$$
$$y(t) = \Phi(t), \qquad \text{for} \quad t \le t_0$$
$$y'(t) = \Phi'(t), \qquad \text{for} \quad t \le t_0. \tag{6.4}$$

Here both $\Phi(t)$ and $\Phi'(t)$ are necessary to provide the history of dependent variables and their derivatives before the start of the simulation. While $\tau$ and $\sigma$ are the (constant) delays, the functions $\alpha_i(t, y(t))$ and $\beta_i(t, y(t))$ are more general delay functions.

## 6.2 Difficulties when Solving DDEs

There are several difficulties associated with DDEs that one does not have with ODEs. We deal with two of those in the next sections.

### 6.2.1 Discontinuities in DDEs

One prominent difference between DDEs and ODEs is the necessity of the initial functions ($\Phi$, $\Phi'$) to be specified. These replace the initial conditions specified for an ODE and which are given as just one vector of values specified *at* the initial point of the integration interval. In contrast, the initial functions provide the lagged values and derivatives that extend back *before* the initial point of the integration.

A first problem when solving DDEs is that, as the initial functions are generally not fully compatible with the rest of the model solution, a *discontinuity* may emerge as the solver switches from the initial function to past values saved during the integration. This happens when, at the initial point, the right-hand derivative $y'(t_0)^+$ is not equal to the left-hand derivative $\Phi'(t_0)$, or when $\Phi(t)$ has discontinuities.

Moreover, these discontinuities will propagate in time, i.e. for constant time-delays $\tau$ or $\sigma$, they will re-appear at integer multiples of $\tau$ or $\sigma$. In the case where the delays are defined in the dependent variables only, the discontinuity will be smoothed as it will occur in successively higher derivatives. However, in the case where the DDE also includes delays in the derivatives (neutral DDEs), the discontinuities will persist. This is important to realise as the standard solvers assume that the solution is sufficiently continuous over an integration step. Thus, the existence of discontinuities may provoke numerical difficulties, requiring the solver to take very small steps in the vicinity of discontinuities.

We demonstrate the discontinuities that may arise in DDEs by considering two very simple delay differential equations, solved in the interval $t = [0, 10]$. The code for solving these equations will be given in the next chapter (Chap. 7).

In the first DDE example (6.5), the derivative of the solution at $t$ depends on the *value* of the solution at $(t - 1)$. The history function for times preceding the initial time ($t \in [-1, 0]$) is taken to be equal to 1:

$$\begin{aligned} y' &= -y(t-1), \\ y(t) &= 1, \quad \text{for} \quad t \in [-1, 0]. \end{aligned} \tag{6.5}$$

It is clear that the left and right-hand derivatives do not agree for $t = 0$ since $\Phi'(0) = 0$ ($y$ is a constant) and this is $\neq y'(0)^+ = -1$.

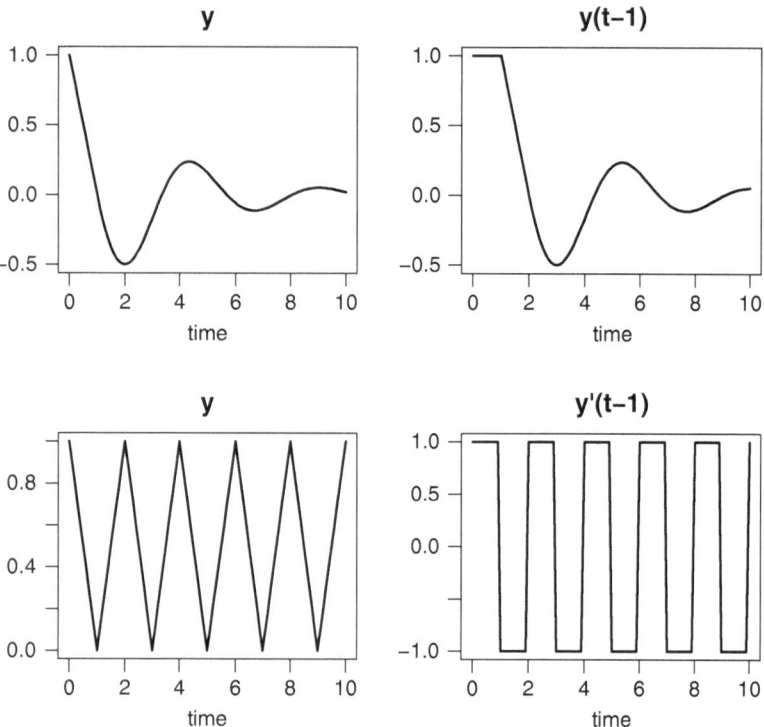

**Fig. 6.1** Two simple delay differential equations; *above*: DDE with delays in the dependent variables only, see (6.5), *below*: neutral DDE, see (6.6). Note how in the first example, the output is smooth, while it is highly discontinuous in the latter

In the second, neutral case (6.6), the derivative of the solution at $t$ depends on the *derivative* at $(t-1)$. The history function for the derivative is taken to be equal to 1.

$$
\begin{aligned}
y' &= -y'(t-1), \\
y'(t) &= 1, \qquad \text{for} \quad t \in [-1,0].
\end{aligned}
\tag{6.6}
$$

Similarly as to what we had in the previous DDE case, here also the right- and left-hand derivatives at $t = 0$ do not correspond, as $\Phi'(0) = 1$ while $y'(0)^+ = -1$.

The solution to these two delay differential equations (6.5) and (6.6) are plotted in Fig. 6.1. In the first case (6.5), the solution is a damped oscillation, slightly discontinuous at $t = 1$, and increasingly less so at $t = 2, 3$, etc. ... (Fig. 6.1 upper). The solution of the neutral DDE (6.6) in contrast, remains equally discontinuous at these times (Fig. 6.1 lower).

### 6.2.2   Small and Vanishing Delays

A second problem when solving DDEs is that a delay may vanish, i.e. $\sigma \to 0$ or $\tau \to 0$. When the delays approach zero, the problem is called a *vanishing delay* problem.

These are generally much harder to solve than the other DDEs. This is because, as the delays become smaller, they may become smaller than the current time step and, for explicit codes, this will require extrapolation. A strategy to deal with vanishing delays in implicit methods is described in [7].

## 6.3 Numerical Methods for Solving DDEs

A survey by [2] shows that the lag functions that arise most frequently in the literature are constants. Indeed problems with constant delays are a large and important class of problem, and by confining our attention to these problems particularly efficient solvers can be derived. In particular, solvers based on explicit Runge-Kutta formulae are often especially efficient for these problems.

A popular approach to the implementation of DDEs is to simply extend ODE codes with functions to retrieve past values and derivatives. These simple DDE solvers just keep track of previous values and derivatives of the dependent variables, and add functions that calculate the values and derivatives at requested previous points. They leave the treatment of the discontinuities to be dealt with by the numerical integrator. This is the approach developed in the solver dede in R . For an ODE code to be suitable for this task, it has to possess a "dense output" facility, i.e. the possibility to estimate the values of dependent variables and their derivatives at arbitrary points in the integration interval. This is necessary in order to evaluate past solutions and their derivatives at points that were not mesh points, and hence not saved. For instance, when using a Runge-Kutta method the coefficients of the interpolating polynomial used for dense output is saved at each integration step (see Sect. 2.1.2.5). To retrieve past values one applies the dense output formula to the coefficients embracing the requested integration point. When based on linear multistep methods an accurate interpolant at the requested lagged time can be obtained either via cubic Hermite interpolation [10] or using the dense output formulae, based on the Nordsieck history array (see Sect. 2.2.3.1).

There also exist more complex codes that take special action at discontinuities, e.g. RADAR5 [7], DDE_SOLVER [12] and DDEM [14]. These implementations of DDE solution methods can solve much harder problems than the strategy used in R . Other effective codes for solving DDEs are ARCHI [11], DKLAG6 [5], and DDVERK [6].

## References

1. Baker, C. T. H., Paul, C. A. H., & Wille, D. R. (1995). *A bibliography on the numerical solution of delay differential equations* (Numerical Analysis Rep. No. 269). Manchester, UK.
2. Baker, C. T. H., Paul, C. A. H., & Wille, D. R. (1995). Issues in the numerical solution of evolutionary delay differential equations. *Advances in Computational Mathematics, 3*, 171–196.

3. Bellen, A., & Zennaro, M. (2003). *Numerical methods for delay differential equations. Numerical mathematics and scientific computation.* New York: The Clarendon Press/Oxford University Press.
4. Bellen, A., Maset, S., Zennaro, M., & Guglielmi, N. (2009). Recent trends in the numerical solution of retarded functional differential equations. *Acta Numerica, 18*, 1–110.
5. Corwin, S. P., Sarafyan, D., & Thompson, S. (1997). **DKLAG6**: A code based on continuously imbedded sixth order Runge–Kutta methods for the solution of state dependent functional differential equations. *Applied Numerical Mathematics, 24*, 319–333.
6. Enright, W. H., & Hayashi, A. (1997). A delay differential equation solver based on a continuous Runge–Kutta method with defect control. *Numerical Algorithms, 16*, 349–364.
7. Guglielmi, N., & Hairer, E. (2008). Computing breaking points in implicit delay differential equations. *Advances in Computational Mathematics, 3*, 229–247.
8. Hale, J. K., Verduyn, L., & Sjoerd, M. (1993). *Introduction to functional-differential equations: Vol. 99. Applied mathematical sciences.* New York: Springer
9. May, R. M. (1975). *Stability and complexity in model ecosystems* (2nd ed.). Princeton: Princeton University Press.
10. Oberle, H. J., & Pesch, H. J. (1981). Numerical treatment of delay differential equations by hermite interpolation. *Numerische Mathematik, 37*, 235–255.
11. Paul, C. A. H. (1995). *A user-guide to* **ARCHI** (Numerical Analysis Rep. No. 283). University of Manchester, Manchester.
12. Thompson, S., & Shampine, L. F. (2006). A friendly fortran DDE solver. *Applied Numerical Mathematics, 53*(3), 503–516.
13. Tziperman, E., Stone, L., Cane, M. A., & Jarosh, H. (1994). El nino chaos: Overlapping of resonances between the seasonal cycle and the pacific ocean-atmosphere oscillator. *Science, 264*, 72–74.
14. Zivari, H., & Enright, W. H. (2010). An efficient unified approach for the numerical solution of delay differential equations. *Numerical Algorithms, 53*(2–3), 397–417.

# Chapter 7
# Solving Delay Differential Equations in R

**Abstract** DDEs are solved in R much in the same way as ODEs, i.e. by the multistep methods from the R package **deSolve** that are used to solve initial value problems of ODEs. One major difference between DDEs and initial value problems for ODEs is the presence of a "memory" term which retrieves past values of the dependent variable or of the derivatives. Two functions are provided that retrieve past values and derivatives.

## 7.1 Delay Differential Equation Solvers in R

A popular approach to implement DDEs is to extend ODE codes with functions to retrieve past values and derivatives, and this is the approach adopted in the R package **deSolve** [11].

Function dede from the R package **deSolve** can solve delay differential equations, while the past values and past derivatives are available via function lagvalue and lagderivs respectively. A simplified syntax of these functions is:

```
dede (y, times, func, parms, method, ...)
lagvalue (t, nr)
lagderiv (t, nr)
```

Function dede is similar to ode (see Sect. 3.1). Arguments t and nr of functions lagvalue and lagderiv are the time point at which the value/derivative value is required, while nr is the position of the variable. If nr is not specified, the values or derivatives of all dependent variables will be returned.

The various integration methods that can be used to solve DDEs in R can be found in the appendix (Sect. A.3, Table A.8).

K. Soetaert et al., *Solving Differential Equations in R*, Use R!,
DOI 10.1007/978-3-642-28070-2_7, © Springer-Verlag Berlin Heidelberg 2012

## 7.2  Two Simple Examples

We first implement the two simple problems (6.5) and (6.6) from (Sect. 6.2.1).

### 7.2.1  DDEs Involving Solution Delay Terms

In the first DDE example (7.1), the derivative of the solution at $t$ depends on the *value* of the solution at $t - 1$. The history function for times preceding the initial time ($t = [-1, 0]$) sets the past value equal to 1. The equation is:

$$y' = -y(t-1),$$
$$y(t) = 1, \quad \text{for} \quad t \in [-1, 0]. \tag{7.1}$$

In the R implementation (function DDE1) the if-else statement provides either access to the history value (if tlag preceeds the initial time) or to the saved past values (if tlag > 0). The function lagvalue extracts the past value at the requested time point.

```
library(deSolve)
DDE1 <- function(t, y, parms) {
    tlag <- t - 1
    if (tlag <= 0)
       ylag <- 1
    else
       ylag <- lagvalue(tlag)

    list(dy = - ylag, ylag = ylag)
  }
yinit <- 1
times <- seq(from = 0, to = 10, by = 0.1)
yout  <- dede(y = yinit, times = times, func = DDE1,
              parms = NULL)
```

The solution of this first model is $1 - t$ for $0 \le t \le 1$, while it is $t^2/2 - 2t + 3/2$ for $1 \le t \le 2$, so we can compare the numerical solution, generated with dede with these analytic values. The maximal deviation is $1.4e{-}6$ (not shown). We can make the solution more precise by setting the relative and absolute tolerances to a lower value. For instance, adding the arguments (atol = 1e-10, rtol= 1e-10) to the call to dede reduces the deviation with the analytic solution to $1.4e{-}10$.

### 7.2.2  DDEs Involving Derivative Delay Terms

The implementation of the simple neutral DDE from (Sect. 6.2.1) in R is very similar to the previous DDE example except that the past values of the *derivatives* are obtained by a call to R function lagderiv. The equations are:

$$y' = -y'(t-1),$$
$$y'(t) = 1, \qquad \text{for} \quad t \in [-1,0].$$

(7.2)

They are implemented in R as:

```
DDE2 <- function(t, y, parms) {
    tlag <- t - 1
    if (tlag <= 0)
      ylag <- 1
    else
      ylag <- lagderiv(tlag)

    list(dy = - ylag, ylag = ylag)
}
yout2 <- dede(y = yinit, times = times, func = DDE2,
              parms = NULL)
```

The output of these two models can be found in Fig. 6.1 from the previous chapter.

## 7.3  Chaotic Production of White Blood Cells

We now implement a more realistic example. A well-studied delay differential equation model is the Mackey-Glass equation [5], which models the production of white blood cells and is given by:

$$y' = ay_\tau \frac{1}{1+y_\tau^c} - by,$$
$$y_\tau = y(t-\tau),$$
$$y_t = 0.5, \qquad \text{for} \quad t \le 0.$$

(7.3)

Here $y$ is the current density of the circulating white blood cells, $y_\tau$ is the density $\tau$ time-units in the past, $b$ is the destruction rate. The first term in the equation for $y'$ is the introduction of new blood cells in the blood, in response to the demand at a previous time ($\tau$). The value of $(1+y_\tau^c)^{-1}$ is almost 1 for small values of $y_\tau$ and decreases towards near-zero values for large values of $y_\tau$. The larger the value of the parameter $c$, the steeper the decline. We use $c = 10$ here. Thus, blood cell production will be near its maximal rate ($a$) if past densities are small.

This equation generates chaotic solutions for certain values of the parameters, such as $a = 0.2$, $b = 0.1$, $\tau = 20$.

The implementation of this model in R is simple. As we will run the model for different values of $\tau$, we pass the value of tau to the solver which will include it in its call to the derivative function (mackey).

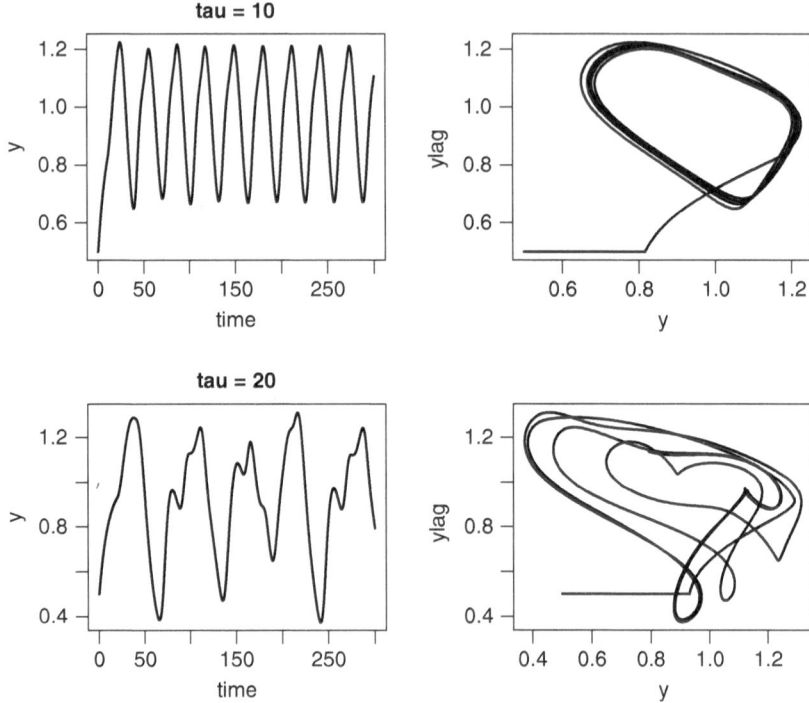

**Fig. 7.1** The Mackey-Glass delay differential equation, solved with parameter $\tau = 10$ (*above*) and $\tau = 20$ (*below*). See text for the R code

```
library(deSolve)
mackey <- function(t, y, parms, tau) {
    tlag <- t - tau
    if (tlag <= 0)
      ylag <- 0.5
    else
      ylag <- lagvalue(tlag)
    dy <- 0.2 * ylag * 1/(1+ylag^10) - 0.1 * y
    list(dy = dy, ylag = ylag)
  }
```

```
yinit <- 0.5
times <- seq(from = 0, to = 300, by = 0.1)
```

```
yout1 <- dede(y = yinit, times = times, func = mackey,
              parms = NULL, tau = 10)
yout2 <- dede(y = yinit, times = times, func = mackey,
              parms = NULL, tau = 20)
```

When solved with $\tau = 10$, the output is periodic, while cell densities display a chaotic pattern for $\tau = 20$ (Fig. 7.1)

```
plot(yout1, lwd = 2, main = "tau = 10",
     ylab = "y", mfrow = c(2, 2), which = 1)
plot(yout1[,-1], type = "l", lwd = 2, xlab = "y")
plot(yout2, lwd = 2, main = "tau = 20",
     ylab = "y", mfrow = NULL, which = 1)
plot(yout2[,-1], type = "l", lwd = 2, xlab = "y")
```

## 7.4   A DDE Involving a Root Function

Similarly to what happens in the case of ODEs, the solver will have some difficulties at discontinuities, so it may be beneficial to report to the solver when the discontinuities arise and then reinitialise the integration.

Below is a DDE with discontinuities in the derivatives, the Mariott-Delisle Controller Problem [6]. This example was used in [9] to show how to use a root function together with DDE solvers. The DDE involves a step function in the history term ($\Delta$):

$$y' = (-y(t) + \pi(a + \varepsilon\, sign(\Delta) - u\sin^2(\Delta)))/\tau,$$
$$\Delta = y(t - 12) - x_b, \qquad\qquad (7.4)$$
$$y_t = 0.6, \qquad \text{for} \quad t \le 0,$$

for $x_b = -0.427$, $a = 0.16$, $\varepsilon = 0.02$, $u = 0.5$, $\tau = 1$. The term $sign(\Delta)$ causes a sharp change in the dynamics as it switches from $+1$ to $-1$ and vice versa. The implementation in R is:

```
xb  <- -0.427; a <- 0.16; xi <- 0.02; u <- 0.5; tau <- 1
yinit <- c(y = 0.6)
```

```
mariott <- function(t, y, parms) {
   tlag <- t - 12
   if (tlag <= 0)
     ylag <- 0.6
   else
     ylag <- lagvalue(tlag)

   Delt  <- ylag - xb
   sDelt <- sign(Delt)

   dy <- (-y + pi*(a + xi*sDelt - u*(sin(Delt))^2))/tau
   list(dy)
 }
times <- seq(from = 0, to = 120, by = 0.5)
yout <- dede(y = yinit, times = times, func = mariott,
             parms = NULL)
```

When letting the solver track the discontinuities, this problem is solved in 3,282 function evaluations. We can see this by requesting the "diagnostics" of the solution (diagnostics (yout)).

It is more robust to let the solver locate the change in sign of $\Delta$ (Delt), which occurs at a root in $\Delta$. Thus, we provide a root and event function (see Sects. 3.4.2 and 3.4.3). The root function is to locate where the value of Delt changes sign, the event function ensures that the solution proceeds unchanged after that.

```
root <- function(t, y, parms) {
   tlag <- t - 12
   if (tlag <= 0)
     return (1) # not a root
   else
     return(lagvalue(tlag)- xb)
 }
event <- function(t, y, parms) return(y)
```

```
yout <- dede(y = yinit, times = times, func = mariott,
             parms = NULL, rootfun = root,
             events = list(func = event, root = TRUE))
```

This solution method is only slightly more efficient (3,161 function evaluations). A plus however is that it has recorded the times when the root was found:

```
attributes(yout)$troot
```

```
[1]   14.01588   24.49263   67.54677   75.18143 118.43615
```

We add the root positions as grey vertical lines to the plot (Fig. 7.2):

```
plot(yout, lwd = 2,
     main = "Mariott-Delisle Controller problem")
abline(v = attributes(yout)$troot, col = "grey")
```

## 7.5   Vanishing Time Delays

While there are many DDEs that dede *can* solve, there are also some that *cannot* easily be solved with the current implementation of the DDE solution method in R. These are problems belonging to the category of vanishing time delay differential equations, for which the step size chosen by the solver exceeds the time delay requested. In the way that delays were implemented in R, the solver then needs to *extrapolate* rather than interpolate the variable's value or derivative to the requested time point. Only quite sophisticated algorithms [3] deal with this in a robust way, and this is not (yet) included in R. There are however some vanishing delay problems that can be solved with dede, although not without difficulties.

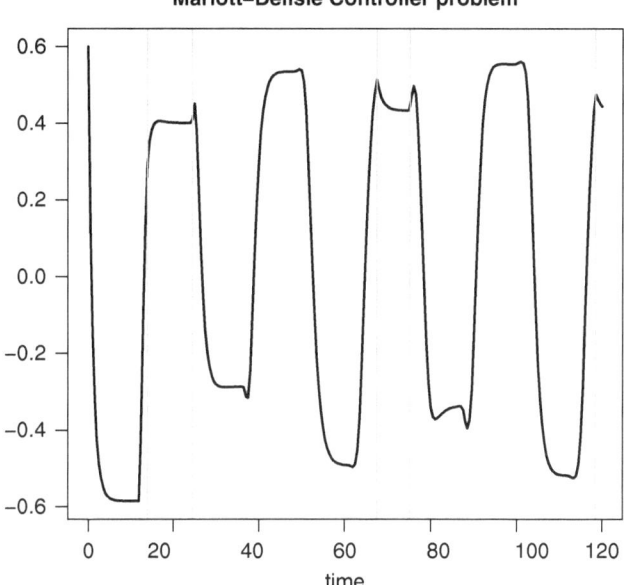

**Fig. 7.2** Solution of the Mariott-Delisle controller problem; the *grey lines* are the positions of the root. See text for the R code

We implement the following problem from [3], which was modified from [2]:

$$y' = \cos(t)(1 + y(ty^2(t))) + cy(t)y'(ty^2(t))$$

$$+ (1 - c)\sin t \cos(t \sin^2 t) - \sin(t + t \sin^2 t), \qquad (7.5)$$

$$y(t) = y'(t) = 0, \qquad \text{for} \quad t <= 0.$$

This problem contains delays in the dependent variables and in the derivatives and has a vanishing delay at $t = 0, \pi/2, 3\pi/2,\ldots$ Notwithstanding the complexity of this formula, it has $y(t) - \sin(t)$ as exact solution!

Below is an implementation of this problem, with an estimate of the error. This model cannot be solved by R for $c = 1$. We increase the precision of the solution by setting `atol` and `rtol` to a very low value.

```
vanishing <- function(t, y, parms, cc) {
    tlag <- t*y^2
    if (tlag <= 0) {
        ylag  <- 0
        dylag <- 0
    } else {
        ylag  <- lagvalue(tlag)
        dylag <- lagderiv(tlag)
    }
```

```
   dy <- cos(t)*(1+ylag) + cc*y*dylag +
        (1-cc)*sin(t)*cos(t*sin(t)^2) - sin(t+t*sin(t)^2)

   list(dy)
  }
yinit <- c(y = 0)
times <- seq(from = 0, to = 2*pi, by = 0.1)
yout <- dede(y = 0, times = times, func = vanishing,
             parms = NULL,cc = -0.5, atol = 1e-10, rtol = 1e-10)
```

We estimate the maximal error it has generated:

```
print(max(abs(yout[,2] - sin(yout[,1]))))
```

```
[1] 1.81828e-06
```

## 7.6  Predator-Prey Dynamics with Harvesting

In Sect. 3.4 we showed how to include events and roots in ordinary differential
equations. Systems of DDEs can also be subject to impulses. Here we include an
example, from fisheries, [1], to show how we can use dede to include events that
are triggered by a state-dependent condition. As the times at which the events occur
are not known in advance, this is a rather difficult problem to solve.

The model describes predator-prey dynamics where the prey density ($N$) is
regulated by its density at a previous time $t - \tau_1$, and with carrying capacity $K$.
The prey is preyed upon at a per capita rate $aH(t)$ and this is converted into
predator density ($H$) with efficiency $b$. The predator development time $\tau_2$ is taken
into account in the predator dynamics. Predators die at a rate equal to $d$,

$$N' = rN(t)\left(1 - \frac{N(t-\tau_1)}{K}\right) - aN(t)H(t)$$

$$H' = abN(t-\tau_2)H(t-\tau_2) - dH(t). \tag{7.6}$$

Implemented in R this becomes:

```
LVdede <- function(t, y, p) {
  if (t > tau1) Lag1 <- lagvalue(t - tau1) else Lag1 <- yini
  if (t > tau2) Lag2 <- lagvalue(t - tau2) else Lag2 <- yini

  dy1 <- r * y[1] *(1 - Lag1[1]/K) - a*y[1]*y[2]
  dy2 <- a * b * Lag2[1]*Lag2[2] - d*y[2]

  list(c(dy1, dy2))
  }
```

These differential equations (7.6) apply as long as the prey's density ($N$) does not exceed a critical level, $Ycrit$. At the critical density $Ycrit$, they are harvested by fishermen, which reduces their density by 30%.

Thus, when $N = Ycrit$, the dynamics of prey is governed by the "event"

$$N(t)^+ = 0.7N(t)^-. \tag{7.7}$$

In R we define a root function that will locate the times at which this critical density is reached (rootfun), while the reduction of prey density is performed in an event function (eventfun). This function must return both dependent variables, i.e. including the predator density (y[2]), although this does not change.

```
rootfun <- function(t, y, p)
   return(y[1] - Ycrit)
eventfun <- function(t, y, p)
   return (c(y[1] * 0.7, y[2]))
```

Function dede is used to solve the delay differential equation. We print the first ten times at which the critical prey density was reached (troot).

```
r <- 1; K <- 1; a <- 2; b <- 1; d <- 1; Ycrit <- 1.2*d/(a*b)
tau1 <- 0.2; tau2 <- 0.2
yini <- c(y1 = 0.2, y2 = 0.1)
times <- seq(from = 0, to = 200, by = 0.01)
yout <- dede(func = LVdede, y = yini, times = times,
             parms = 0, rootfun = rootfun,
             events = list(func = eventfun, root = TRUE))
attributes(yout)$troot [1:10]
```

```
[1]    2.125283   3.057600   3.991063   4.926748   5.864435
[6]    6.803803   7.745041   8.688137   9.632305  10.578815
```

The output (Fig. 7.3) shows how the predators ($y_2$) first decrease gradually while the prey ($y_1$) initially increase until they reach the critical density ($Ycrit = 0.6$). After harvesting, the prey density is instantaneously reduced and predator density initially decreases to increase after that, concurrently with prey density, until they are set back again. This continues for a while; each time the predator density is slightly higher, until the prey density does not reach the critical level anymore and the predator – prey density converges to their equilibrium values.

```
plot(yout[,-1], type = "l")
```

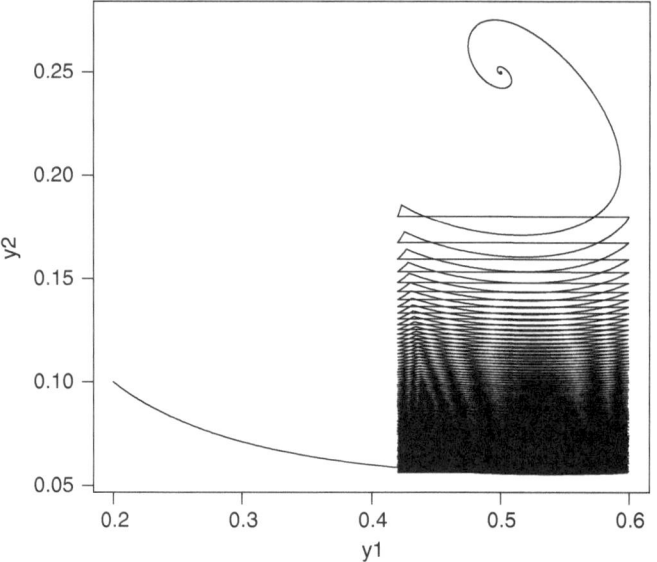

**Fig. 7.3** Solution of the fisheries model, a DDE including events. See text for the R code

## 7.7  Exercises

### 7.7.1  The Lemming Model

A nice variant of the logistic model from Sect. 3.1.1, is the DDE lemming model as from [9], citing [12].

$$y' = ry\left(1 - \frac{y(t-\tau)}{K}\right),\tag{7.8}$$

with initial condition $y(t=0) = 19.001$, and parameter values $r = 3.5$, $\tau = 0.74$, $K = 19$, and history $y(t) = 19$ for $t < 0$.

Try to recreate Fig. 7.4:

### 7.7.2  Oberle and Pesch

Implement the problem described by [7]

$$y' = -\lambda y(t-1)(1 + y(t)),\tag{7.9}$$

on the interval [0, 20] and with history $y(t) = t$ for $t \leq 0$, for four values of the parameter $\lambda$; 1.5, 2.0, 2.5 and 3.

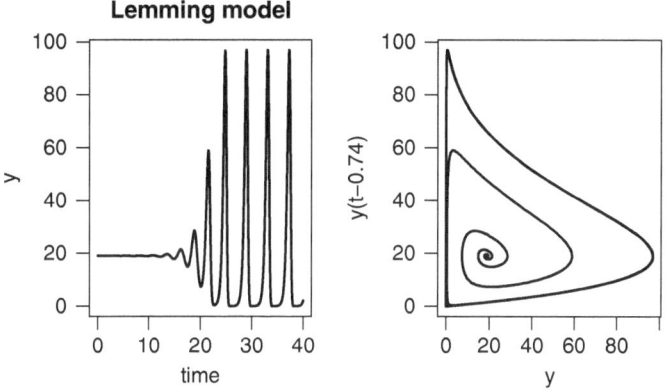

**Fig. 7.4** The lemming model

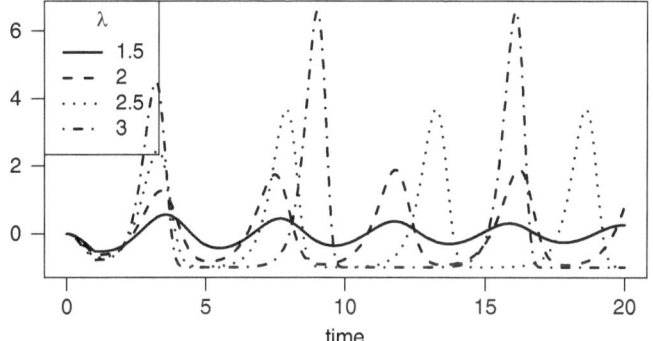

**Fig. 7.5** The Oberle and Pesch model

Try to recreate Fig. 7.5.

### 7.7.3  An Epidemiological Model

The following example is from [4]. The system of DDEs describes the progress of an epidemic by a SIR model (Susceptible, Infected, Recovered). The reference solution at $t = 40$ is: $(0.0912491205663460, 0.0202995003350707, 5.98845137909849)$. The equations are:

$$y_1' = -y_1(t)y_2(t-1) + y_2(t-10)$$
$$y_2' = y_1(t)y_2(t-1) - y_2(t)$$
$$y_3' = y_2(t) - y_2(t-10),$$

(7.10)

for t ∈ [0,40] and

$$Y(t) = [5, 0.1, 1]^T, \qquad t \leq 0. \tag{7.11}$$

Try to solve the problem on a mesh of 400 points.

### 7.7.4 A Neutral DDE

The following neutral DDE [8]:

$$\begin{aligned} y' &= y(t) + y'(t-1), & t \geq 0 \\ y(t) &= 1, & t < 0, \end{aligned} \tag{7.12}$$

has as analytic solution: $y(t) = e^t$ for $t \in [0, 1]$, $y(t) = (t-1)e^{(t-1)} + e^t$ for $t \in [1, 2]$. Try to solve the problem on a mesh of 100 points.

### 7.7.5 Delayed Cellular Neural Networks With Impulses

Another problem from [1] describes delayed cellular neural networks with impulsive effects. This DDE model exhibits both discrete and continuous behavior over the time interval of interest. The discrete jumps in the states (events) occur at particular points in time. As the delays for this problem at times vanish during the integration, this is a relatively difficult problem. The relatively complex equations are:

$$\begin{aligned} y_1' &= -6y_1 + \sin(2t)f(y_1) + \cos(3t)f(y_2) \\ &\quad + \sin(3t)f(y_1(\tau_1)) + \sin(t)f(y_2(\tau_2)) + 4\sin(t) \\ y_2' &= -7y_2 + \cos(t)f(y_1)/3 + \cos(2t)f(y_2)/2 \\ &\quad + \cos(t)f(y_1(\tau_1)) + \cos(2t)f(y_2(\tau_2)) + 2\cos(t) \\ f(x) &= 0.5(|x+1| - |x-1|) \\ \tau_1 &= t - 0.5(1 + \cos(t)) \\ \tau_2 &= t - 0.5(1 + \sin(t)), \end{aligned} \tag{7.13}$$

on the interval [0,40] and with history $y(t) = (-0.5, 0.5)$ for $t \leq 0$. Every two time units, the dependent variables are increased, $y_1$ by 20%, $y_3$ by 30%. You will need to write an "event-function" to make these changes, while passing a vector of times at which the events are to take place.

The results are in Fig. 7.6, try to recreate it. This problem was solved in [10].

**Fig. 7.6** The DDE with
impulses example

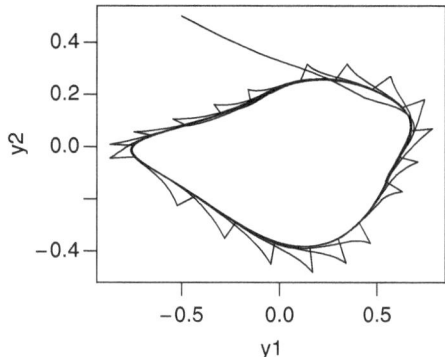

# References

1. Corwin, S. P., Thompson, S., & White, S. M. (2008). Solving ODEs and DDEs with impulses. *Journal of Numerical Analysis, Industrial and Applied Mathematics (JNAIAM), 3*(1–2), 139–149.
2. Enright, W. H., & Hayashi, A. (1998). A delay differential equation solver based on a continuous Runge–Kutta method with defect control. *Numerical Algorithms, 16*, 349–364.
3. Guglielmi, N., & Hairer, E. (2008). Computing breaking points in implicit delay differential equations. *Advances in Computational Mathematics, 29*(3), 229–247.
4. Hairer, E., & Wanner, G. (1996). *Solving ordinary differential equations II: Stiff and differential-algebraic problems.* Heidelberg: Springer.
5. Mackey, M. C., & Glass, L. (1977). Oscillation and chaos in physiological control systems. *Science, 197*, 287–289.
6. Marriott, C., & DeLisle, C. (1989). Effects of discontinuities in the behavior of a delay differential equation. *Physica D, 36*, 198–206.
7. Oberle, H. J., & Pesch, H. J. (1981). Numerical treatment of delay differential equations by hermite interpolation. *Numerische Mathematik, 37*, 235–255.
8. Paul, C. A. H. (1994). *A test set of functional differential equations* (Numerical Analysis Rep. No. 243). University of Manchester, Manchester.
9. Shampine, L. F., & Thompson, S. (2001). Solving DDEs in MATLAB. *Applied Numerical Mathematics, 37*, 441–458.
10. Soetaert, K., & Petzoldt, T. Solving ODEs, DDEs, DAEs, and PDEs in the open source software R. *Journal of Numerical Analysis, Industrial and Applied Mathematics (JNAIAM)*.
11. Soetaert, K., Petzoldt, T., & Setzer, R. W. (2010). Solving differential equations in R: Package **deSolve**. *Journal of Statistical Software, 33*(9), 1–25.
12. Tavernini, L. (1996). *Continuous-time modeling and simulation.* Amsterdam: Gordon and Breach.

# Chapter 8
# Partial Differential Equations

**Abstract** A characteristic of partial differential equations (PDEs) is that the solution changes as a function of more than one independent variable. Usually these variables are time and one or more spatial coordinates. The numerical solution of a PDE therefore often requires the solution to be approximated not only in time as in ODEs, but in space as well. If *all* derivatives are approximated by finite differences at a finite number of points, a set of algebraic equations is obtained whose solution can be found using root solving algorithms. This is the common approach for solving time-independent PDEs. In contrast, PDEs which involve time as one of the independent variables are usually solved with the method of lines. In this case only *spatial* derivatives are discretised, while the *time* derivative is left as a continuous function. The result is a system of ODEs in time that can be solved with the initial value problem solvers from previous chapters. Typically, the dimension of the ODE or algebraic system is much larger than the number of components in the original partial differential equation.

## 8.1 Partial Differential Equations

In this chapter we will consider the numerical solution of PDEs using finite difference techniques or the method of lines [13, 15, 23, 31, 32]. There exist other powerful numerical methods for the solution of PDEs such as finite element, finite volume or meshless methods [18, 21, 24], but since they are not implemented in the R packages (finite elements, meshless) or we give no examples here (finite volume), they will not be discussed in this chapter.

In partial differential equations the solution changes as a function of more than one independent variable, often time, and one or more spatial variables. As a simple example, consider the advection-diffusion equation:

$$\frac{\partial Y}{\partial t} = -v\frac{\partial Y}{\partial x} + D\frac{\partial^2 Y}{\partial x^2}, \tag{8.1}$$

K. Soetaert et al., *Solving Differential Equations in R*, Use R!,
DOI 10.1007/978-3-642-28070-2_8, © Springer-Verlag Berlin Heidelberg 2012

where $t$ denotes time, $x$ is the spatial position, $v$ is the advection rate, $D$ is a positive constant known as the diffusion coefficient, and $Y$ is the dependent variable. As $Y$ depends on both $x$ and $t$, the equation is called a partial differential equation; $\partial Y/\partial t$ is the (first order) partial derivative of $Y$ with respect to $t$, $\partial^2 Y/\partial x^2$ is the (second order) partial derivative with respect to $x$. As the highest derivative is 2, the PDE is said to be a second order equation.

A PDE that contains only a diffusion term belongs to a fundamental class of PDEs called a *parabolic* PDE. Another well-known PDE is the wave equation, which is of *hyperbolic* type, e.g:

$$\frac{\partial^2 u}{\partial t^2} = c^2 \frac{\partial^2 u}{\partial x^2}. \tag{8.2}$$

A third basic PDE is the *elliptic* PDE, such as the 2-dimensional Poisson equation:

$$\frac{\partial^2 w}{\partial x^2} + \frac{\partial^2 w}{\partial y^2} = f(x,y). \tag{8.3}$$

Examples of these three types will be given in the next chapter.

### 8.1.1   Alternative Formulations

There exist several alternative notations for dealing with PDEs. Sometimes partial differentiation is denoted using subscripts, e.g.

$$u_x = \frac{\partial u}{\partial x}$$

$$u_{xx} = \frac{\partial^2 u}{\partial x^2} \tag{8.4}$$

$$\partial_x = \frac{\partial}{\partial x}.$$

Another more compact representation of derivatives is provided by coordinate-independent operators. For example, the symbol $\nabla$ (nabla) represents the *gradient operator*, which defines the vector field, consisting of the partial derivatives:

$$\nabla = \frac{\partial}{\partial x}$$

$$\nabla = \left( \frac{\partial}{\partial x}, \frac{\partial}{\partial y} \right)^T \tag{8.5}$$

$$\nabla = \left( \frac{\partial}{\partial x}, \frac{\partial}{\partial y}, \frac{\partial}{\partial z} \right)^T,$$

in 1-D, 2-D and 3-D cartesian coordinates respectively (see below for other coordinate systems).

Related to the gradient operator is the *divergence operator*, $\nabla\cdot$. If we denote by $\underline{a} = (a_1, a_2, a_3)^T$, a vector $\in \mathfrak{R}^3$, then the divergence of $\underline{a}$ is given by:

$$\nabla \cdot \underline{a} = \left( \frac{\partial a_1}{\partial x} + \frac{\partial a_2}{\partial y} + \frac{\partial a_3}{\partial z} \right). \tag{8.6}$$

It is also common to use the *Laplace operator*, $\Delta$, a second order differential operator defined as the divergence of the gradient. If $u$ is a differentiable scalar function in $\mathfrak{R}^3$:

$$\Delta u = \nabla \cdot \nabla u = \nabla^2 u = \frac{\partial^2 u}{\partial x^2} + \frac{\partial^2 u}{\partial y^2} + \frac{\partial^2 u}{\partial z^2}. \tag{8.7}$$

When using these formalisms, the advection-diffusion equation (8.1) can be written as:

$$\begin{aligned}
\frac{\partial Y}{\partial t} &= -v\frac{\partial Y}{\partial x} + D\frac{\partial^2 Y}{\partial x^2} \\
Y_t &= -vY_x + DY_{xx} \\
\frac{\partial Y}{\partial t} &= -v\nabla \cdot Y + D\Delta Y,
\end{aligned} \tag{8.8}$$

where the last equation is also valid for other coordinate systems or for multi-dimensional (2-D, 3-D) problems. We can also write the equation such that it applies to spatially variable $D$ and $v$:

$$\frac{\partial Y}{\partial t} = -\nabla \cdot (vY) + \nabla \cdot (D\nabla Y). \tag{8.9}$$

If we now rewrite the last equation as:

$$\frac{\partial Y}{\partial t} = -\nabla \cdot (vY - D\nabla Y), \tag{8.10}$$

the equation is written in the form of a conservation law and we obtain the *flux-conservative* representation of the equation. Large classes of time-dependent PDEs can be put in flux-conservative form. This is more generally represented as:

$$\frac{\partial Y}{\partial t} = -\nabla \cdot F. \tag{8.11}$$

The equation states that the change in the quantity $Y$ with time ($t$) equals the negative divergence of the flux $F$. In (8.10) the advective flux is represented as a velocity $v$ times the quantity, while the diffusive flux equals a diffusion coefficient $D$ times the negative of the quantities gradient. As we will see later, this formula is the basis of the functions for solving partial differential equations in the R package **ReacTran** [33] (see Chap. 9).

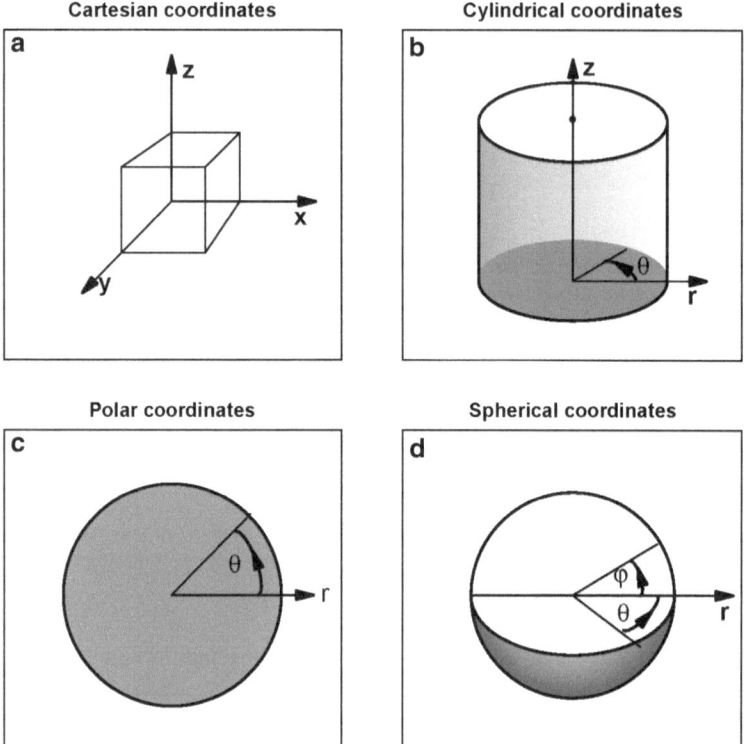

**Fig. 8.1**  Different coordinate systems

## 8.1.2   Polar, Cylindrical and Spherical Coordinates

In the previous section we used *cartesian* coordinates to specify the position of quantities along the $x$, $y$, and $z$ direction (Fig. 8.1a). However, many problems are more easily specified in one of the alternative coordinate systems, such as the polar, cylindrical or spherical coordinate system.

For instance, substances dispersing on a flat surface from a point source are most conveniently represented in *polar* coordinates, which have a circular boundary (Fig. 8.1c). A natural extension of polar coordinates to the z-axis gives way to the *cylindrical* coordinate system (Fig. 8.1b). This representation is for instance very well suited to describe heat transport in a cylindrical tube. If a quantity is radiating in all spatial directions from a point source, then the *spherical* coordinate system is generally best (Fig. 8.1d).

Whereas in three-dimensional cartesian coordinates $(x, y, z)$, we write:

$$\nabla u = \left( \frac{\partial u}{\partial x}, \frac{\partial u}{\partial y}, \frac{\partial u}{\partial z} \right)^T$$

$$\nabla \cdot \underline{a} = \frac{\partial a_1}{\partial x} + \frac{\partial a_2}{\partial y} + \frac{\partial a_3}{\partial z} \tag{8.12}$$

$$\Delta f = \frac{\partial^2 f}{\partial x^2} + \frac{\partial^2 f}{\partial y^2} + \frac{\partial^2 f}{\partial z^2},$$

in the cylindrical coordinate system $(r, \theta, z)$[1] this becomes (Fig. 8.1b):

$$\nabla u = \left( \frac{\partial u}{\partial r}, \frac{1}{r}\frac{\partial u}{\partial \theta}, \frac{\partial u}{\partial z} \right)^T$$

$$\nabla \cdot \underline{a} = \frac{1}{r}\frac{\partial}{\partial r}(a_1 r) + \frac{1}{r}\frac{\partial a_2}{\partial \theta} + \frac{\partial a_3}{\partial z} \tag{8.13}$$

$$\Delta f = \frac{1}{r}\frac{\partial}{\partial r}\left( r\frac{\partial f}{\partial r} \right) + \frac{1}{r^2}\frac{\partial^2 f}{\partial \theta^2} + \frac{\partial^2 f}{\partial z^2},$$

and in spherical coordinates $(r, \theta, \varphi)$[2] (Fig. 8.1d) we have:

$$\nabla u = \left( \frac{\partial u}{\partial r}, \frac{1}{r}\frac{\partial u}{\partial \theta}, \frac{1}{r\sin\theta}\frac{\partial u}{\partial \varphi} \right)^T$$

$$\nabla \cdot \underline{a} = \frac{1}{r^2}\frac{\partial}{\partial r}(r^2 a_1) + \frac{1}{r\sin\theta}\frac{\partial}{\partial \theta}(\sin\theta\, a_2) + \frac{1}{r\sin\theta}\frac{\partial a_3}{\partial \varphi} \tag{8.14}$$

$$\Delta f = \frac{1}{r^2}\frac{\partial}{\partial r}\left( r^2\frac{\partial f}{\partial r} \right) + \frac{1}{r^2\sin\theta}\frac{\partial}{\partial \theta}\left( \sin\theta\frac{\partial f}{\partial \theta} \right) + \frac{1}{r^2\sin^2\theta}\frac{\partial^2 f}{\partial \varphi^2}.$$

The polar coordinate system $(r, \theta)$[3] (Fig. 8.1c) is a two-dimensional set of coordinates similar to the cylindrical system, but without the $z$-dependence. It describes properties distributed on a circle:

$$\Delta f = \frac{1}{r}\frac{\partial}{\partial r}\left( r\frac{\partial f}{\partial r} \right) + \frac{1}{r^2}\frac{\partial^2 f}{\partial \theta^2}. \tag{8.15}$$

### 8.1.3 Boundary Conditions

Just as we needed *initial conditions* to fully specify an initial value problem for an ODE, partial differential equations are only complete after we have defined what happens on the *boundaries* of the domain. For spatial coordinates, the boundaries are where the model interacts with the outside world. The number of boundary

---

[1] Sometimes denoted by $(r, \varphi, z)$.

[2] Sometimes denoted by $(r, \varphi, \theta)$.

[3] Sometimes denoted by $(r, \varphi)$.

conditions needed in an independent variable equals the highest order derivative in this variable.

Often, the model domain is represented by the symbol $\Omega$ while the boundaries are denoted by $\partial\Omega$.

The boundary conditions can be represented as:

$$\alpha Y + \beta \frac{\partial Y}{\partial n} = g \quad \text{on} \quad \partial\Omega, \tag{8.16}$$

where $\partial Y/\partial n$ denotes the derivative normal to the boundary and $\alpha$ and $\beta$ are constants. Several types of boundary conditions are commonly used:

- If $\alpha = 0$, the gradient is specified at the boundaries and the problem is said to have a *Neumann* boundary condition.
- With $\beta = 0$, the variable values are specified and the boundary is referred to as being of *Dirichlet* type.
- If both $\alpha$ and $\beta$ differ from 0, we have a *Robin* problem.
- Another frequently used boundary condition is the *periodic* boundary condition, where a variable's value and/or its gradient is equal at opposite ends of the boundary.

## 8.2   Solving PDEs

The solution of PDEs using a finite difference approach or the method of lines involves several steps. First, one or more continuous independent variables are subdivided in a number of grid cells, and the continuous derivatives are replaced by discrete, algebraic approximate equations.

For time-varying cases, it is customary to discretise the spatial coordinate(s) only, while time is left in continuous form. This is called the *method of lines*, and in this way, one PDE is translated into a large number of coupled ordinary differential equations that can be solved using standard initial value problem solvers. For time-independent problems, usually all independent variables are discretised and all derivatives approximated by algebraic equations, which are then solved by root-finding techniques.

In general the dimension of the resulting ODE or algebraic system is much larger than the number of components in the original PDE.

There are two ways in which to reduce the numerical errors associated with the numerical differencing. The simplest way is to increase the number of grid points, but this increases the system of algebraic equations or ODEs that will need to be solved. When using the method of lines more gridpoints also increase the stiffness of the problem. An alternative is to resort to higher order difference schemes.

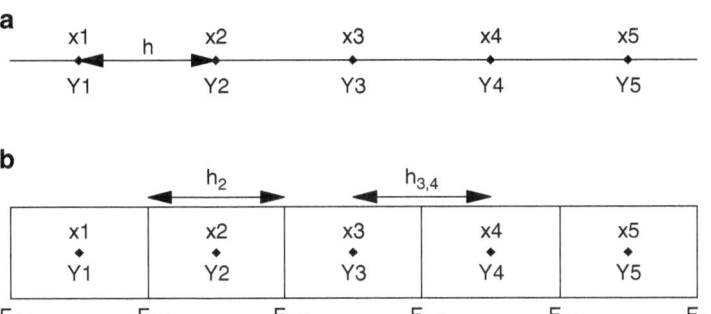

**Fig. 8.2** Discretisation of the spatial domain. (**a**) Simple differencing grid. (**b**) Staggered grid

## 8.3 Discretising Derivatives

Given that PDEs almost never have known solutions, they need to be solved by estimating the values of variables at a limited number of points in the model domain only. For simplicity, we explain the technique in one dimension. It is straightforward to extend the numerical approximations to 2-D and 3-D model domains, by discretising at a regular grid of points, such that the values of the variables are represented in a matrix (2-D) or an array (3-D).

Assume that the spatial domain is $\Omega = [a, b]$ and that a large number of points are spaced regularly on a grid ($\Omega_h = a = x_0, x_1, \ldots, x_N, b = x_{N+1}$), and with distance between two points equal to $h$ (i.e. $x_i = x_{i-1} + h$).

The variable *values*, $Y(t,x)$, are defined at the node points $x_i$ only, and the approximation of $Y(t,x_i)$ at $x_i$ is denoted by $Y_i(t)$ (Fig. 8.2.). Even though this representation is not continuous, if we consider a large enough number of values, we will have a good idea of how the variable changes in the model domain.

The differential equations also include the *derivatives* of the variable, which we need to represent. Since we only know the variable values $Y_i$ at selected points $x_i$, we use these to represent the derivatives, using difference quotients. Thus, the $n$th derivative of $Y$ at the point $x_i$ is estimated as:

$$Y_i^{(n)}(t) = \frac{1}{h^n} \sum_{k=-r}^{s} c_k Y_{i+k}(t), \qquad (8.17)$$

with suitable coefficients $c_k$ [15]. In this formula we use $r$ values to the left and $s$ values to the right of $x_i$, such that a total of $r+s+1$ values $c_k$ need to be determined. The set of points $x_{i-r}, \ldots, x_{i+s}$ is called the *stencil* of the discretisation around $x_i$ [15].

As was the case for the development of suitable Runge-Kutta methods (see Sect. 2.1.2) a number of order conditions should be met for the difference equation (8.17) to have a certain order of accuracy. Moreover, these equations are often used to transport concentrations or densities, and an important condition for these quantities is that they cannot become negative.

A useful property of solutions to certain PDEs is that they should satisfy the maximum principle [10, 23]. In particular the Maximum Principle is concerned with properties of solutions of certain partial differential equations of elliptic and parabolic type and basically it says that the maximum of a function in a domain is to be found on the boundary of that domain.

In what follows we will discuss suitable spatial discretization schemes for the first and second order derivatives, for which we will use the advection and diffusion equation respectively. To preserve the analytic properties of the solution we should use only numerical approximations that also conserve non-negativity. Unfortunately, this strongly restricts the order of the difference schemes that can be used. We will see in the next sections that, if non-negativity needs to be ensured, the order of the simple linear schemes cannot exceed 1 or 2 for first order (advection) and second order (diffusion) derivatives respectively. For many practical cases, the first order discretisation of advection is not adequate, and non-linear discretisations are used. This is dealt with in Sect. 8.3.4.

### 8.3.1   Basic Diffusion Schemes

Consider the diffusion equation in one spatial dimension ($x$):

$$\frac{\partial Y(t,x)}{\partial t} = D\frac{\partial^2 Y(t,x)}{\partial x^2},\tag{8.18}$$

with initial condition $Y(0,x) = g_0(x)$ and boundary conditions of Dirichlet type

$$\begin{aligned}Y_{t,x=a} &= g_a\\Y_{t,x=b} &= g_b.\end{aligned}\tag{8.19}$$

Applying (8.17) to the right-hand side of (8.18), the general spatial finite difference scheme for the $i$th component of this equation which is second order in space, can be written as:

$$Y_i'(t) = \frac{D}{h^2}\sum_{k=-r}^{s} c_k Y_{i+k}(t), \quad i = 1,\ldots N,\tag{8.20}$$

with suitable coefficients $c_k$. As diffusion is direction independent, in general $s$ is taken to be equal to $r$, and $c_k = c_{-k}$.

The conditions to obtain an order $q$ are $\sum_k c_k = 0$, $\sum_k k^2 c_k = 2$, $\sum_k k^4 c_k = 0,\ldots$, $\sum_k k^q c_k = 0$ [15, p. 65], which can be satisfied for $q = 2s$.

For example, with $s = 1$ and $s = 2$, we obtain the *second order* and *fourth order central diffusion discretisation* respectively [15, p. 62, 65]:

$$Y_i'(t) = \frac{D}{h^2}(Y_{i-1}(t) - 2Y_i(t) + Y_{i+1}(t)) \tag{8.21}$$

$$Y_i'(t) = \frac{D}{h^2}(-\frac{1}{12}Y_{i-2}(t) + \frac{4}{3}Y_{i-1}(t) - \frac{5}{2}Y_i(t) + \frac{4}{3}Y_{i+1}(t) - \frac{1}{12}Y_{i+2}(t)). \tag{8.22}$$

At the boundaries the fourth order central diffusion discretisation should be used with a suitable discretization scheme for $Y_1'(t)$ and $Y_N'(t)$ to avoid the use of values of $x_i$ outside the domain $\Omega$.

The discretization of the equations generates an initial value problem which can be represented as:

$$Y'(t) = AY(t) + b, \qquad t > 0$$
$$Y(t = 0) = (g_0(x_1), \dots, g_0(x_N))^T, \tag{8.23}$$

where $Y(t) = (Y_1(t), \dots, Y_N(t))^T$, $A$ is a banded matrix whose elements depend on the coefficients of the discretization and $b$ depends on the boundary conditions. For example, if we use the second order discretization (8.21), $A$ is a tridiagonal matrix with $-2D/h^2$ on the main diagonal and $D/h^2$ on the upper and lower diagonals and $b = (Dg_a/h^2, 0, \dots, 0, Dg_b/h^2)^T$.

#### 8.3.1.1 Non-negativity of Diffusion Schemes

In order for a formula to maintain non-negativity, it is required that the resulting IVP (8.23) maintains non-negativity. A necessary and sufficient condition for (8.23) to have a positive solution, if the initial condition is positive, is that the matrix $-A$ is an M–matrix [22, p. 206] (see [30] for the definition of M–matrices). This means that it is required that $Dc_k \geq 0$ for all $k \neq 0$ and $Dc_0 < 0$. Clearly, this is met for the second order formula (8.21), but due to the negative coefficients $c_{-2} = c_2 = -1/12$ the fourth order discretisation (8.22) does not conserve non-negativity.

Indeed, for the general diffusion discretisation (8.20), the requirement for non-negativity leads to an order barrier of 2 [15, p. 120], i.e. it is not possible to derive non-negative formulae of order higher than 2. Fortunately, for many practical applications, second order approximations are sufficiently accurate. Naturally, to have a full non-negative scheme, the numerical method used to solve the IVP (8.23) should preserve this property.

### 8.3.2 Basic Advection Schemes

Following the approach that we used when dealing with diffusion, we can also approximate the advection equation in one spatial dimension ($x$):

$$\frac{\partial Y(t,x)}{\partial t} = -v\frac{\partial Y(t,x)}{\partial x}, \tag{8.24}$$

with initial condition $Y(t = 0, x) = g_0(x)$ and boundary conditions of Dirichlet type $Y(t, x = a) = g_a$. Based on (8.17), the general spatial finite difference scheme for the $i$th component of this first order equation can be written as:

$$Y_i'(t) = -\frac{v}{h} \sum_{k=-r}^{s} c_k Y_{i+k}(t), \quad i = 1, \ldots N, \tag{8.25}$$

generating an ordinary differential equation of the same type as (8.23).

### 8.3.2.1  Stability

Based on [15, p. 59], the conditions needed to obtain a scheme of order $q$ are $\sum_k c_k = 0$, $\sum_k k c_k = -1$, $\sum_k k^2 c_k = 0, \ldots, \sum_k k^q c_k = 0$, and it is possible to find a unique set of coefficients $c_{-r}, \ldots, c_s$ which gives order $q = r + s$, for any $r$ and $s$.

However, a fundamental result obtained by [17] is that if $v > 0$ then the scheme of order $q = r + s$ is stable only for $s \leq r \leq s + 2$, while for $v < 0$, the condition is $r \leq s \leq r + 2$.

For instance, the *first order upwind* difference approximation uses only two points to approximate the first derivative, and is different for positive or negative $v$ [15, p. 53]:

$$\begin{aligned} Y_i'(t) &= -v\frac{Y_i(t) - Y_{i-1}(t)}{h} \quad v > 0 \\ Y_i'(t) &= -v\frac{Y_{i+1}(t) - Y_i(t)}{h} \quad v < 0. \end{aligned} \tag{8.26}$$

In this case the matrix $A$ of the IVP (8.23) generated after the discretization is a bidiagonal matrix with, for positive $v$, $-v/h$ on the main diagonal and $v/h$ on the first lower diagonal and the vector $b$ in (8.23) is $b = (v/h)(g_a, 0, \ldots, 0, 0)^T$. It is easy to show that this IVP is stable (see Sect. 1.2.4).

Another simple approximation of the advection equation is the *second order central* difference approximation, which, as it is symmetric, remains the same for positive or negative $v$ [15, p. 53]:

$$Y_i'(t) = -v\frac{Y_{i+1}(t) - Y_{i-1}(t)}{2h}. \tag{8.27}$$

A higher order example is the *third order upwind-biased* scheme [15] where $r = 2$ and $s = 1$:

$$\begin{aligned} Y_i'(t) &= \frac{v}{h}(-\frac{1}{6}Y_{i-2}(t) + Y_{i-1}(t) - \frac{1}{2}Y_i(t) - \frac{1}{3}Y_{i+1}(t)) \quad v > 0 \\ Y_i'(t) &= \frac{v}{h}(\frac{1}{3}Y_{i-1}(t) + \frac{1}{2}Y_i(t) - Y_{i+1}(t) + \frac{1}{6}Y_{i+2}(t)) \quad v < 0. \end{aligned} \tag{8.28}$$

Near the boundaries, the *third order upwind-biased* scheme should be used with a suitable discretization scheme for $Y_1'(t)$ to avoid the use of an $x_i$ value outside the boundary domain.

### 8.3.2.2  Non-negativity of Advection Schemes

The IVP arising after the discretization of the space derivative (8.23) maintains non-negativity if the matrix $-A$ is an M-matrix (see Sect. 8.3.1.1). This requires that $vc_k \geq 0$ for all $k \neq 0$ and $vc_0 < 0$. Whereas this is clearly satisfied for the first order upwind scheme (8.26), this is neither the case for the second order central difference scheme (8.27) nor for the third order upwind-biased scheme (8.28).

Indeed, there exists an order barrier of *one* for maintaining non-negativity of advection schemes ([9] [15, p. 119]). This means that all schemes of order higher than 1 do not ensure that the dependent variables will remain positive or zero.

As the first order approximation has some undesirable properties, other approximation formulae for the advection equation have been devised to attain higher accuracy while still conserving non-negativity. These more complex advection schemes will be dealt with in Sect. 8.3.4.

## 8.3.3  Flux-Conservative Discretisations

The approximations from the previous sections work well if the parameters, such as the diffusion coefficient $(D)$ and velocity $(v)$ are constant, and for a constant grid size. When applied carelessly, they may however lead to (numerical) mass creation or loss in the case where these properties change with spatial position, or for irregular grid sizes. In such circumstances, it is simplest to discretise the partial differential equations in flux-conservative form (8.10), which for the 1-D advection-diffusion equation (8.1) can be written as:

$$\frac{\partial Y}{\partial t} = -\frac{\partial F}{\partial x}$$
$$F = v(x) \cdot Y - D(x) \cdot \frac{\partial Y}{\partial x}, \tag{8.29}$$

where $F$ is the (advective-diffusive) flux.

The discretisation of this PDE comprises the introduction of a staggered finite-difference grid. Here the dependent variable values $Y_i$ are defined in the *centre* of grid cells, while the fluxes $F$ are defined on the grid *interfaces* (Fig. 8.2.)

Denoting by $F_{i,i+1}$ the flux on the interface from cell $i$, to cell $i+1$ we can derive the following approximation to (8.29):

$$\frac{dY_i}{dt} = -\frac{F_{i,i+1} - F_{i-1,i}}{h_i}$$

$$F_{i,i+1} = v_{i,i+1}Y_i - D_{i,i+1}\frac{Y_{i+1} - Y_i}{h_{i,i+1}}, \qquad (8.30)$$

which applies for positive $v$ and where we have used centered differences for representing the diffusive term and the first order upwind approximation for the advective term. In this equation, $h_i$ is the width of grid cell $i$, while $h_{i,i+1}$ is the distance from the middle of grid cell $i$ to grid cell $i+1$ (Fig. 8.2b). As the flux over the boundaries is used in the derivatives of both adjacent cells mass will be conserved.

### 8.3.4  More Complex Advection Schemes

The advection equation, despite its apparent simplicity, is a very difficult process to approximate numerically with a reasonable accuracy. As seen in Sect. 8.3.2, this is because advection schemes of order higher than one may produce negative quantities, so they cannot be used if quantities should remain non-negative. However, linear schemes of order one have very low accuracy. We first illustrate the severity of the problem associated with using first order advection schemes, based on a well-known test example, and then give some remedies.

#### 8.3.4.1  Advection of a Square Pulse

In Fig. 8.3 the simple advection equation

$$\frac{\partial Y}{\partial t} = -v\frac{\partial Y}{\partial x}, \qquad (8.31)$$

is approximated using several standard approximation schemes, which are the first order upwind (8.26), the third order upwind-biased (8.28), and the second order central advection (8.27) discretizations respectively.

In the test application, a steep pulse is set in motion with a constant velocity. As there is no production or consumption of the quantity $Y$ the pulse should retain its shape, but clearly (Fig. 8.3) this is not so for any of the schemes tested!

The simplest method, the first order upwind scheme causes an unrealistic flattening of the solution, as if there was a substantial dispersion in addition to the advection. This dispersion-like behavior leads this artifact to be called "numerical dispersion" (Fig. 8.3a). Using central difference schemes to approximate advection retains the shape of the pulse better, but these schemes give numerical solutions with large oscillations (Fig. 8.3c), and negative values. These oscillations are even more pronounced when using first order forward downwinded differences (not shown).

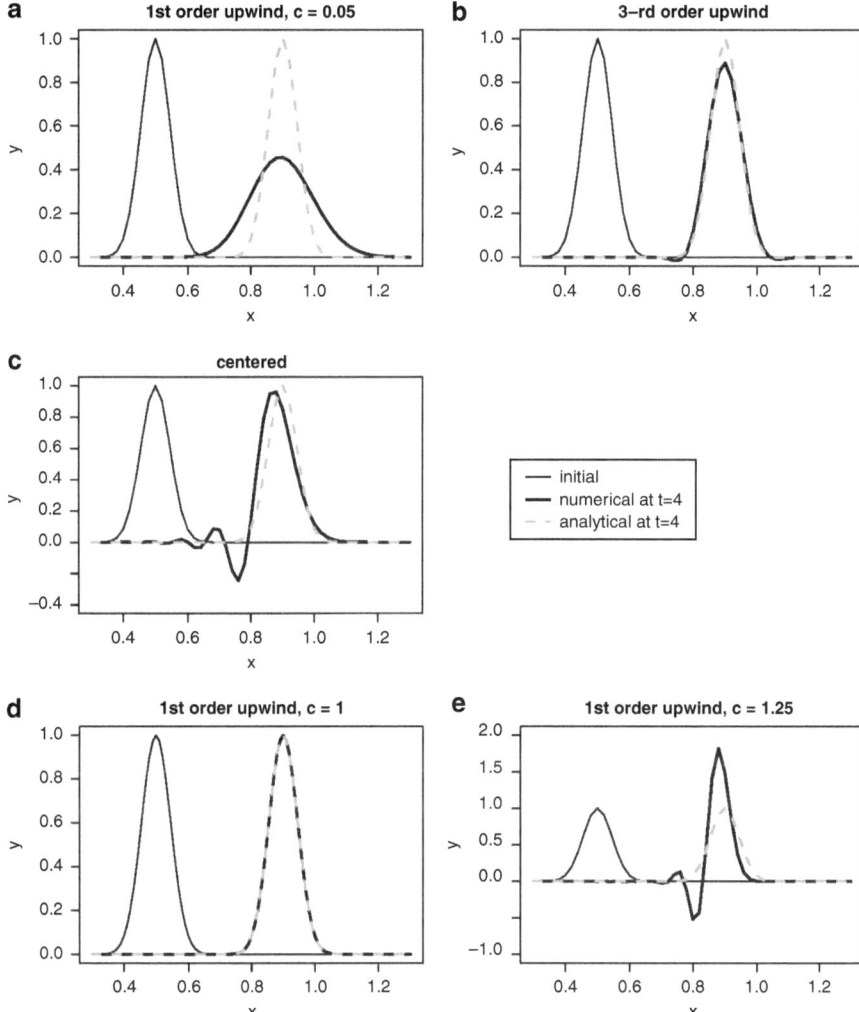

**Fig. 8.3** The advection equation, approximated on a domain with 50 grid points, with $h = 0.02$. The equations are solved with the Euler method, using a time step of 0.1, and using different advection schemes; the velocity $= 0.1$. (**a**) First order upwind scheme. (**b**) The third order upwind-biased scheme. (**c**) Second order central differences. The Courant number in (**a**–**c**) equals 0.05. (**d**) First order upwind scheme with a Courant number of 1, and (**e**) first order upwind scheme with a Courant number of 1.25. The *dark, thin line* is the initial condition, the *dashed line* is the exact solution at $t = 4$, the *thick line* is the numerical approximation

A more realistic profile is given by the third order upwind-biased discretisation, but this still shows some oscillations and small negative values (Fig. 8.3b). Of the three methods tested, only the first order upwind scheme did not produce negative values. This undesirable result can be dealt with in several ways.

### 8.3.4.2   Time Step Conditions

An important property when differencing advection terms, especially when using explicit numerical integration methods is the *speed* at which quantities propagate in the difference equation. If the physical velocity $v$ equals the algorithm speed $h/\Delta t$ then the numerical approximation will be exact. However, if it exceeds the algorithm speed, the scheme is unstable. For the Euler method used with the first order upwind scheme, this leads to the following *Courant* condition for these methods to be stable:

$$|v| \leq \frac{h}{\Delta t}, \tag{8.32}$$

more commonly written as:

$$\frac{|v|\Delta t}{h} \leq 1, \tag{8.33}$$

where the quantity $|v|\Delta t/h$ is called the Courant-Friedricks-Lewy (CFL), or Courant number ($c$) [23]. To avoid instabilities, in many algorithms it is coded that if during a certain time step ($\Delta t$) the maximum Courant number exceeds 1, the time step is split guaranteeing that the split step Courant numbers remain smaller than 1.

In Fig. 8.3a, the Courant number in the upwind scheme was 0.05. The same equation is now solved, again using the upwind scheme, but where the Courant number, $c$, equals 1 (Fig. 8.3d), and 1.25 (Fig. 8.3e). Note the perfect result when $c = 1$, and the numerical oscillations when $c$ exceeds the critical value 1.

Thus, if we have only the advection equation to solve, then when using Euler with a time step such that the Courant number $c = 1$, the result will be exact. Unfortunately, in advection-reaction equations, it is often the reaction part that is the most expensive in terms of computation. Thus, if the equations comprise terms other than advection, simple Euler integration may not work, and we will need to resort to the more complex approximations discussed next.

### 8.3.4.3   Flux Limiters

In order to keep the numerical errors of the advection approximation in check we can use so-called *flux limiters* that ensure positivity and eliminate the numerical oscillations.

For example, total variance diminishing (TVD) schemes [27] ensure that the total variation of the solution of an equation does not increase as time progresses. The principle is as follows. If we write the advective equation in flux-conservative form, we can approximate the flux caused by the advective velocity, $v$, over interface $i, i+1$ using upwind differencing:

$$F_{i,i+1} = v_{i,i+1} Y_i, \tag{8.34}$$

(this formula only applies for $v_{i,i+1} > 0$).

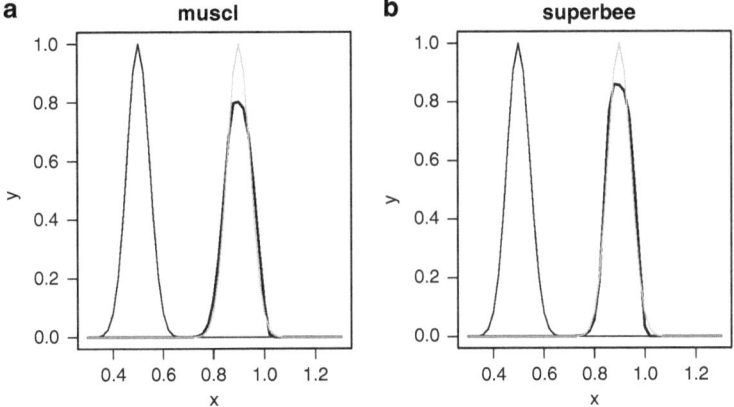

**Fig. 8.4** The advection equation, solved with an upwind-biased scheme, using a flux limiter. (**a**) the muscl scheme and (**b**) the superbee limiter. The *dark, thin line* is the initial condition, the *gray thin line* is the exact solution at t = 4, the *thick line* is the numerical approximation. Note that mass *is* conserved in these limiters, although the figure tends to suggest otherwise

As seen above, this approximation is not very satisfactory, as it produces conspicuous artificial dispersion (Fig. 8.3a). The trick is to add a term that more or less compensates for this numerical artifact while ensuring positivity (e.g. [15, 27]):

$$F_{i,i+1} = v_{i,i+1} \left( Y_i + \Phi_i(\theta_i, c_i) (Y_{i+1} - Y_i) \right). \tag{8.35}$$

Here $\Phi_i$, is called the *limiter function*, and it depends on the slope parameter $\theta_i$ and the Courant number $c_i$. The slope parameter is nonlinear and is given by the ratio:

$$\theta_i = \frac{Y_i - Y_{i-1}}{Y_{i+1} - Y_i}. \tag{8.36}$$

There exist many formulae which can be used as limiter functions. Two commonly used methods are the *superbee* [28] and the *muscl* [34] limiter:

$$\Phi_i = \frac{1}{2}(1 - c)(\max(0, \min(1, 2\theta_i), \min(2, \theta_i)))$$

$$\Phi_i = \frac{1}{2}(1 - c)(\max(0, \min(2, (1 + \theta_i)/2, 2\theta_i)). \tag{8.37}$$

In the next figure (Fig. 8.4) we have applied these two limiters to the advection test model. Both produce good results, although superbee tends to cause a steepening of the smooth function, while muscl is slightly more diffusive than superbee.

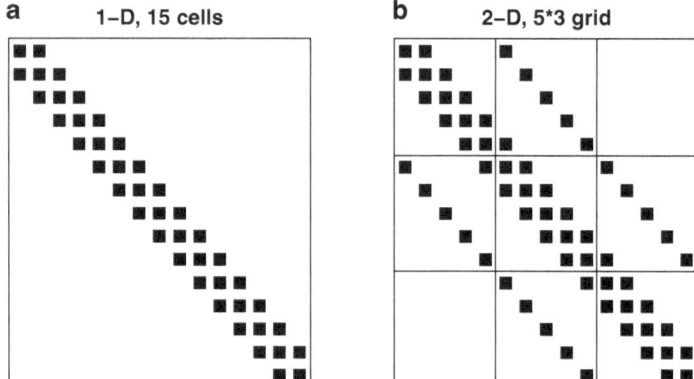

**Fig. 8.5** The pattern of non-zero elements in the Jacobian for a 1-D and 2-D problem, approximated using centered differences

## 8.4 The Method Of Lines

In the previous sections we focussed on the approximation of the spatial derivatives. We still need to find a solution to the PDE. For time-varying equations it is customary to discretise the spatial coordinate(s) only while time is left in continuous form. In this way, the partial differential equation is no longer a continuous function of space, but has become a set of coupled ordinary differential equations (such as (8.23)). The resulting ODEs are of rather large size, in the 1-D case N times the number of dependent variables, where N is the number of grid cells. They can be solved with the standard integration methods as described in previous chapters.

The technique of discretising spatial derivatives only, while solving the time derivative with an IVP solver is called the *method of lines* (MOL). The name refers to the solution that consists of values at each spatial grid point, and along time "lines".

Of course, as the continuous PDEs are replaced by approximate differences, numerical errors are introduced. These can generally be controlled by refining the grid (i.e. reducing $h$), but this makes the IVP more stiff, larger in size and therefore more difficult to solve by an IVP solver. Many of these equations are therefore best solved with efficient implicit methods, such as BDF or MEBDFI (Sect. 2.5). However, as they are generally large systems of equations, this may pose quite a computational challenge!

Fortunately, the spatial discretisation leads to ODEs with a Jacobian whose non-zero elements are restricted to a small number of bands, parallel to the diagonal. An example of the pattern of the Jacobian, for a problem describing advective-diffusive transport of one component, in one dimension and in two dimensions is in Fig. 8.5. A large reduction in computation time is then achieved by using an integrator adapted to efficiently solve such sparse systems.

Several ODE or DAE codes provide methods to efficiently solve sparse systems. For instance, the codes LSODA [26], LSODE [14], VODE [3], DASSL [25], DASPK [2], MEBDFI [6], RADAU5 [12], BIMD [5] and GAMD [16] all include linear solvers based on banded matrices [7] and can therefore efficiently solve 1-D problems. The code LSODES [14] uses the Yale sparse matrix solver [8] and is very well suited for solving 2-D or 3-D problems. Apart from the direct (dense or banded) linear solution methods, the codes DASPK [2], VODPK [3] and LSODPK [14] also include several preconditioned Krylov subspace iterative methods. This makes them ideally suited to solve very large (1-D, 2-D or 3-D) problems [4]. A set of PDE integration routines based on the method of lines and written in widely accepted and used languages can be found in [11, 20].

## 8.5   The Finite Difference Method

For time-invariant problems, usually all independent variables are discretised, and all derivatives approximated by algebraic equations. A root-solving method is then used to solve the large and sparse set of algebraic equations. The same efficiency can be attained by also considering the sparseness structure of the Jacobian when solving these algebraic systems. A classical Newton-Raphson method that makes use of banded [1] or arbitrarily sparse matrices is most efficient to solve 1-D or multi dimensional problems respectively. Apart from the direct sparse matrix solvers such as in [8] also the iterative solution methods (e.g. [29, 30]) are useful.

This classical *finite difference method* can also be used to discretise both spatial and temporal derivatives [10]. A well-known example is the Lax-Wendroff scheme for advection ([19] [15, p. 98]).

Note that there exists another powerful method to discretise the spatial domain, the finite element method [18]. However, as this method is not yet available in R, we do not discuss it here.

## References

1. Anderson, E., Bai, Z., Bischof, C., Demmel, J., Dongarra, J., Du Croz, J., Greenbaum, A., Hammarling, S., McKenney, A., Ostrouchov, S., & Sorensen, D. (1992). **LAPACK**'s *user's guide*. Philadelphia: Society for Industrial and Applied Mathematics.
2. Brenan, K. E., Campbell, S. L., & Petzold, L. R. (1996). *Numerical solution of initial-value problems in differential-algebraic equations. SIAM classics in applied mathematics.* Philadelphia, PA: Society for Industrial and Applied Mathematics.
3. Brown, P. N., Byrne, G. D., & Hindmarsh, A. C. (1989). **VODE**, a variable-coefficient ODE solver. *SIAM Journal on Scientific and Statistical Computing, 10*, 1038–1051.
4. Brown, P. N., Hindmarsh, A. C., & Petzold, L. R. (1994). Using krylov methods in the solution of large-scale differential-algebraic systems. *SIAM Journal on Scientific and Statistical Computing, 15*(6), 1467–1488.

5. Brugnano, L., Magherini, C., & Mugnai, F. (2006). Blended implicit methods for the numerical solution of DAE problems. *Journal of Computational and Applied Mathematics, 189*(1–2), 34–50.

6. Cash, J. R., & Considine, S. (1992). An **MEBDF** code for stiff initial value problems. *ACM Transactions on Mathematical Software, 18*(2), 142–158.

7. Dongarra, J. J., Bunch, J. R., Moler, C. B., & Stewart, G. W. (1979). **LINPACK** *users guide*. Philadelphia, PA: Society for Industrial and Applied Mathematics.

8. Eisenstat, S. C., Gursky, M. C., Schultz, M. H., & Sherman, A. H. (1982). Yale sparse matrix package. I. The symmetric codes. *International Journal for Numerical Methods in Engineering, 18*, 1145–1151.

9. Godunov, S. K. (1959). A finite difference method for the numerical computation of discontinuous solutions of the equations of fluid dynamics. *Math Sbornik, 47*, 271–306.

10. Greenspan, D., & Casulli, V. (1988). *Numerical analysis for applied mathematics, science, and engineering*. Redwood: Addison-Wesley/Advanced Book Program.

11. Griffiths, G. W., & Schiesser, W. E. (2012). *Traveling wave analysis of partial differential equations. Numerical and analytical methods with MATLAB and MAPLE*. Amsterdam: Elsevier/Academic.

12. Hairer, E., & Wanner, G. (1996). *Solving ordinary differential equations II: Stiff and differential-algebraic problems*. Heidelberg: Springer.

13. Hamdi, S., Schiesser, W. E., & Griffiths, G. W. (2007). Method of lines. *Scholarpedia, 2*(7), 2859.

14. Hindmarsh, A. C. (1983). **ODEPACK**, a systematized collection of ODE solvers. In R. Stepleman (ed.), *Scientific computing: Vol. 1. IMACS transactions on scientific computation* (pp. 55–64). Amsterdam: IMACS/North-Holland.

15. Hundsdorfer, W., & Verwer, J. G. (2003). *Numerical solution of time-dependent advection-diffusion-reaction equations. Springer series in computational mathematics*. Berlin: Springer.

16. Iavernaro, F., & Mazzia, F. (1998). Solving ordinary differential equations by generalized Adams methods: Properties and implementation techniques. *Applied Numerical Mathematics, 28*(2–4), 107–126. Eighth Conference on the Numerical Treatment of Differential Equations (Alexisbad, 1997).

17. Iserles, A., & Strang, G. (1983). The optimal accuracy of difference schemes. *Transactions of the American Mathematical Society, 277*, 779–803.

18. Johnson, C. (2009). *Numerical solution of partial differential equations by the finite element method*. Mineola: Dover. (Reprint of the 1987 edition)

19. Lax, P. D., & Wendroff, B. (1960). Systems of conservation laws. *Communications on Pure and Applied Mathematics, 13*, 217–237.

20. Lee, H. J., & Schiesser, W. E. (2004). *Ordinary and partial differential equation routines in C, C++, Fortran, Java, Maple, and MATLAB*. Boca Raton: Chapman & Hall/CRC.

21. LeVeque, R. J. (2002). *Finite volume methods for hyperbolic problems. Cambridge texts in applied mathematics*. Cambridge: Cambridge University Press.

22. Luenberger, D. G. (1979). *Introduction to dynamic systems: Theory, models, and applications*. New York: Wiley.

23. Mitchell, A. R., & Griffiths, D. F. (1980). *The finite difference method in partial differential equations*. Chichester: Wiley.

24. Pepper, D. (2010). Meshless methods for PDEs. *Scholarpedia, 5*(5), 9838.

25. Petzold, L. R. (1983). A description of **DASSL**: A differential/algebraic system solver. *IMACS Transactions on Scientific Computation*, New Brunswick, NJ, pp. 65–68.

26. Petzold, L. R. (1983). Automatic selection of methods for solving stiff and nonstiff systems of ordinary differential equations. *SIAM Journal on Scientific and Statistical Computing, 4*, 136–148.

27. Pietrzak, J. (1998). The use of TVD limiters for forward-in-time upstream-biased advection schemes in ocean modeling. *Monthly Weather Review, 126*, 812–830.

28. Roe, P. L. (1985). Some contributions to the modeling of discontinuous flows. *Lecture Notes on Applied Mathematics, 22*, 163–193. Amer. Math. Soc., Providence.

29. Saad, Y. (1994). **SPARSKIT**: *A basic tool kit for sparse matrix computations. VERSION 2.*
30. Saad, Y. (2003). *Iterative methods for sparse linear systems* (2nd ed.). Philadelphia: Society for Industrial and Applied Mathematics.
31. Schiesser, W. E. (1991). *The numerical method of lines. Integration of partial differential equations.* San Diego: Academic.
32. Schiesser, W. E., & Griffiths, G. W. (2009). *A compendium of partial differential equation models. Method of lines analysis with MATLAB.* Cambridge: Cambridge University Press.
33. Soetaert, K., & Meysman, F. (2012). Reactive transport in aquatic ecosystems: Rapid model prototyping in the open source software R. *Environmental Modelling and Software, 32,* 49–60.
34. van Leer, B. (1979). Towards the ultimate conservative difference scheme V. A second order sequel to Godunov's method. *Journal of Computational Physics, 32,* 101–136.

# Chapter 9
# Solving Partial Differential Equations in R

**Abstract** R has three packages that are useful for solving partial differential equations. The R package **ReacTran** offers grid generation routines and the discretization of the advective-diffusive transport terms on these grids. In this way, the PDEs are either rewritten as a set of ODEs or as a set of algebraic equations. When solving the PDEs with the method of lines (MOL), the time integration can be performed using specially-designed initial value problem solvers from the R package **deSolve**. When all derivatives have been approximated, functions from the R package **rootSolve** can efficiently solve the algebraic equations. We show how to solve in R the well-known heat equation (parabolic), the wave equation (hyperbolic), Laplace's equation (elliptic), and the advection equation. We then give some more complex examples. Most partial differential equations are defined in cartesian coordinates, but some problems are much better represented in other coordinate systems. These problems can be solved efficiently in R as well.

## 9.1 Methods for Solving PDEs in R

The solution of PDEs basically proceeds in two steps. First a suitable grid is created in one or more of the independent variables, and the equations are numerically approximated on this grid. Then a suitable solution method is used to solve the resulting ODEs or algebraic equations.

### 9.1.1 Numerical Approximations

The R package **ReacTran** [10, 11] offers grid generation routines and the discretization of the diffusive and advective transport terms on these grids.

#### 9.1.1.1  Setting up a Grid

As a first step, a suitable grid can be generated using **ReacTran** functions `setup.grid.1D` and `setup.grid.2D` (there is no corresponding 3D function).

The grid-generating functions are defined as (simplified):

```
setup.grid.1D(x.up, x.down, L, N, ...)
setup.grid.2D(x.grid, y.grid)
```

where `x.up` and `x.down` are the position of the upper and downward boundary, `L` is the total length of the domain, `N` the number of grid points. Two-dimensional grids are specified from two 1-D grids.

#### 9.1.1.2  Numerical Approximation of Advection and Diffusion

**ReacTran**'s transport functions implement finite difference approximations of the diffusion-advection equation on these grids, which for 1-D and in cartesian coordinates is (simplified):

$$-\frac{1}{A} \cdot \frac{\partial}{\partial x} \left[ -A \cdot \left( D \cdot \frac{\partial C}{\partial x} + v \cdot C \right) \right]. \tag{9.1}$$

Here C is the property ("concentration"), $A$ is the (total) surface area ($L^2$) (for most applications $A$ will be 1), and $D$ and $v$ are the diffusion and advection rate respectively.

A simplified form of the syntax for approximating (9.1) in R is:

```
tran.1D(C, C.up, C.down, flux.up, flux.down, D, v,
        A, dx, ...)
```

while for 2-D the corresponding function is:

```
tran.2D(C, C.x.up, C.x.down, C.y.up, C.y.down,
        flux.x.up, flux.x.down, flux.y.up, flux.y.down,
        D.x, D.y, v.x, v.y, A.x, A.y, dx, dy,...)
```

the obvious extension to 3 dimensions for cartesian coordinates, `tran.3D` is not given. These functions implement a flux-conservative discretisation as in (Sect. 8.3.3).

Several kinds of boundary conditions are implemented, including specification of the boundary *values* (`C.up`, `C.down`, `C.x.up`, ...) or *fluxes* (`flux.up`, `flux.down`, `flux.x.up`, ...); fluxes are considered positive in the direction of the axes.

Another **ReacTran** function approximates transport in polar coordinates:

```
tran.polar(C, C.r.up, C.r.down,
           C.theta.up, C.theta.down,
           flux.r.up, flux.r.down,
```

```
            flux.theta.up, flux.theta.down,
            D.r, D.theta, r, theta, ...)
```

while still other functions approximate 3-D transport in cylindrical and spherical coordinates or implement a finite volume method in 1-D (not discussed here).

While these general transport functions also perform advective transport, they include only very rough first order schemes. If higher accuracy for the advective term is required, then it is better to use **ReacTran** function `advection.1D`. This function includes several upstream-biased advection schemes (`adv.method`) containing flux limiters that are based on total variation diminishing concepts, the *superbee* limiter [7], the *muscl* [14] limiter, the *quickest* [5] limiter and the third order upstream-biased polynomial scheme (`adv.method = "super"`, `"muscl"`, `"quick"`, `"p3"` respectively). Their implementation is based on [1]. The syntax for this function is:

```
advection.1D(C, C.up, C.down, flux.up, flux.down, v,
             A, dx, adv.method, ...)
```

## 9.1.2 Solution Methods

When using the MOL, the large and sparse system of ODES can be efficiently solved using specially-designed methods from the package **deSolve** [12]. A simplified form of the syntax for solving 1-D and 2-D PDEs is:

```
ode.1D(y, times, func, parms, dimens, method, ...)
ode.2D(y, times, func, parms, dimens, method, ...)
```

Here `dimens` provides the dimensions, i.e. the number of grid cells in the spatial coordinates, and `method` allows the selection of one of **deSolve's** (**deTestset's**) integration methods. The default method for `ode.1D` is `"lsoda"`. Other good choices are `"lsode"`, `"vode"`, `"daspk"`, `"radau"`, `"bimd"`, `"gamd"` or `"mebdfi"` for stiff problems, or `"adams"`, `"ode45"`, ... for non-stiff problems. For `ode.2D` the default solver used is `"lsodes"`, but for non-stiff problems one of the Runge-Kutta methods or `method = "adams"` will be more efficient.

In the case where all derivatives are approximated, the resulting sparse system of algebraic equations can be conveniently solved with similar methods from the package **rootSolve** [9]:

```
steady.1D(y, func, parms, dimens, method, ...)
steady.2D(y, func, parms, dimens, method, ...)
```

with obvious extension to 3-D. Argument `method` allows us to trigger either the Newton-Raphson technique with banded Jacobian (`method = "stode"`) or with an arbitrarily sparse Jacobian (`method = "stodes"`) or to integrate the model to steady-state (`method = "runsteady"`).

## 9.2   Solving Parabolic, Elliptic and Hyperbolic PDEs in R

In what follows, we first solve very simple examples of three important (geometric) classes of PDEs: parabolic (time-dependent and diffusive), elliptic (time-independent and steady-state), and hyperbolic (time-dependent and wavelike) equations.

### 9.2.1   The Heat Equation

The heat or diffusion equation is the prototype of a *parabolic* partial differential equation. It describes the spreading of heat $(Y)$ in a conductive medium, and is represented by:

$$\frac{\partial Y}{\partial t} = D\Delta Y + r = \nabla \cdot (D\nabla Y) + r, \tag{9.2}$$

where $D$ is a positive quantity, the "diffusion" coefficient, and $r$ is the production rate.

#### 9.2.1.1   Problem Definition

We solve the heat equation on a one-dimensional domain $[0,1]$ with zero production/consumption:

$$\frac{\partial Y}{\partial t} = D\frac{\partial^2 Y}{\partial x^2}. \tag{9.3}$$

For this equation to be fully specified, we need to define the solution on the boundaries of the domain. Since the above one-dimensional equation is second order in space, two boundary values are required. A suitable set of boundary conditions of Dirichlet type is:

$$\begin{aligned} Y_{t,x=0} &= 0 \\ Y_{t,x=1} &= 1. \end{aligned} \tag{9.4}$$

In addition to the boundary conditions, initial conditions must also be specified. This means that, at $t = 0$, the distribution of $Y$ for all $x$ is required. As the equation is first order in time, only one initial condition is needed; we take a sine-wave initial condition.

$$Y(t = 0, x) = \sin(\pi x). \tag{9.5}$$

#### 9.2.1.2   Solving the Heat Equation in R

The solution of (9.3)–(9.5) will give $Y$ as a function of $x$ and $t$. While the discretisation of the time variable, $t$ will be decided upon by the ODE solver (partly influenced by the input argument `times`), we need to explicitly define the spatial discretisation.

Solving this model in R proceeds in several steps. First we use the **ReacTran** function `setup.grid.1D` to subdivide the model domain [0,1] in 100 small, equally-sized cells. Function `setup.grid.1D` generates a staggered grid (see Sect. 8.3.3, Fig. 8.2) and positions the boundary points exactly *on* the boundaries. It returns a list, containing amongst other information the positions at the centre (`x.mid`) and interface (`x.int`) of grid cells , the thicknesses of each grid cell (`dx`) and so on. We assign `x` with the position in the centre of the grid cells, as we will use these values for plotting. We use for the diffusion coefficient, `D.coeff`, a value equal to 0.01.

```
library(ReacTran)
N         <- 100
xgrid     <- setup.grid.1D(x.up = 0, x.down = 1, N = N)
x         <- xgrid$x.mid
D.coeff   <- 0.01
```

The function `Diffusion` describing the ODEs resulting from the MOL is defined as for any initial value problem. It gets as input the current time (`t`), the state variable vector (`Y`, a vector comprising 100 numbers), and the parameters (not used here), and it returns the derivatives `dY` and (optional) other useful quantities.

Within function `Diffusion`, transport is performed by the **ReacTran** function `tran.1D`, where we specify the upstream and downstream boundary conditions (`C.up`, `C.down` respectively, see (9.4)), the diffusion coefficient (`D`) and the spatial grid used (`dx`).

Function `tran.1D` returns a lot of useful information, such as the derivative (`$dC`), and the fluxes across the upstream and downstream boundary (`$flux.up`, `$flux.down`), packed as a `list`. The fluxes are also returned from the function `Diffusion`.

```
Diffusion <- function (t, Y, parms){
    tran <- tran.1D(C = Y, C.up =   0, C.down = 1,
                    D = D.coeff, dx = xgrid)
    list(dY = tran$dC, flux.up = tran$flux.up,
         flux.down = tran$flux.down)
}
```

Next the initial conditions (`Yini`) (see (9.5)) and the `times` at which output is wanted are specified, after which the model is solved, using an appropriate initial value problem solver. As the model is one-dimensional, we use function `ode.1D` here. Function `ode.1D` makes optimal use of the sparsity of the problem; by passing the dimension of the problem (`dimens`), the sparsity pattern can be derived by the solver (see Sect. 8.4). The problem is rather stiff, and therefore the default solver `lsoda`, chosen in `ode.1D` performs quite well. The time it takes to solve the model (in seconds) is printed.

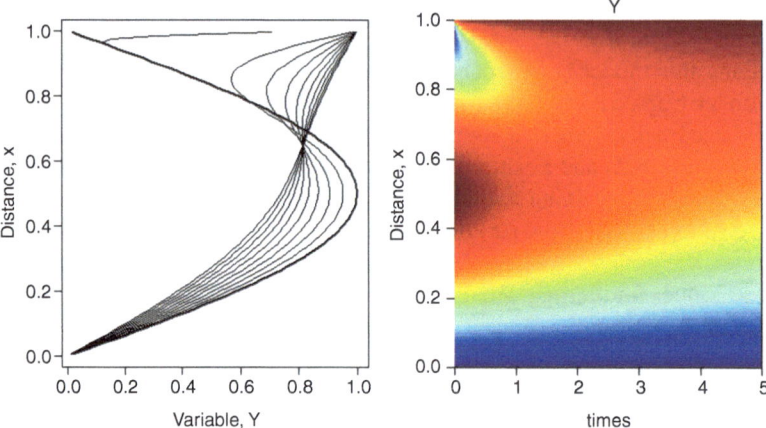

**Fig. 9.1** The solution of the 1-D heat equation, represented as time lines of profiles (*left*) and as a (time, distance) image. The initial condition is the *bold line*. See text for the R code

```
Yini  <- sin(pi*x)
times <- seq(from = 0, to = 5, by = 0.01)
print(system.time(
  out <- ode.1D(y = Yini, times = times, func = Diffusion,
               parms = NULL, dimens = N)
              ))
```

```
  user  system elapsed
  0.36    0.00    0.36
```

We plot the result in two ways, first as a line graph, depicting several profiles at selected times (`plot()` and `lines()`), then as a times-distance `image` (Fig. 9.1). The two figures are arranged in one row, two columns (`mfrow`). While calling **deSolve**'s image function, we ensure that this figure arrangement is not discarded by setting `mfrow = NULL`. Note that, in matrix `out`, the first column has the time, the next N columns contain the profiles (`2: (N+1)`).

```
par (mfrow=c(1, 2))
plot(out[1, 2:(N+1)], x, type = "l", lwd = 2,
     xlab = "Variable, Y", ylab = "Distance, x")
for (i in seq(2, length(times), by = 50))
     lines(out[i, 2:(N+1)], x)
image(out, grid = x, mfrow = NULL, ylab = "Distance, x",
     main = "Y")
```

It is clear from Fig. 9.1 that the imposed value of 1 at the downstream boundary (at $x = 1$), is not consistent with the initial condition (thick line), where this value is 0. So, very rapidly the downstream part of the model takes on the value imposed at the boundary, gradually followed by the internal cells. The steady-state condition, where the heat changes linearly from a value $Y = 1$ downstream to $Y = 0$ upstream is not yet reached at the end of the integration.

## 9.2.2 The Wave Equation

The wave equation is another very important partial differential equation, of the *hyperbolic* type, describing the motion of a wave front. It is used in e.g. acoustic and fluid dynamics and has the following form:

$$\frac{\partial^2 u}{\partial t^2} = \nabla \cdot (c^2 \nabla u), \tag{9.6}$$

where $c$ is the propagation speed of the wave, and $u$ is the variable that changes as the wave passes.

### 9.2.2.1 Problem Definition

We will model the simplest form, a wave $u$, travelling in the $x$-direction only (1-D):

$$\frac{\partial^2 u}{\partial t^2} = c^2 \frac{\partial^2 u}{\partial x^2}. \tag{9.7}$$

This equation describes for instance a pulse traveling through a stretched string. It sets the acceleration of the pulse $(\partial^2 u/\partial t^2)$ proportional to the local curvature $(\partial^2 u/\partial x^2)$, with a factor equal to the square of the speed of the wave $(c)$.

For this equation to have a solution, suitable initial and boundary conditions must be provided. As the equation is second order in time and space, we need two conditions for each one.

We solve the problem with the following initial conditions:

$$u(0,x) = \exp(-\lambda x^2)$$
$$\left. \frac{\partial u}{\partial t} \right|_{0,x} = 0, \tag{9.8}$$

where the first (initial) condition is a Gaussian pulse, and the second (boundary) condition denotes that $u$ starts with zero velocity. The boundary conditions impose a zero value at both ends, located at infinity:

$$u(t, -\infty) = 0$$
$$u(t, \infty) = 0. \tag{9.9}$$

As the analytic solution to this equation is known, we can use this to evaluate the precision of the numerical solution. The analytic solution is:

$$u(t,x) = 0.5(\exp(-\lambda(x + c^2 t)^2) + \exp(-\lambda(x - c^2 t)^2)). \tag{9.10}$$

There are several ways to solve the wave equation. Here we rewrite the second order differential equation as two coupled partial differential equations which are first order in time. This is done by introducing a new variable that represents the first order derivative of variable $u$: $v = \partial u / \partial t$. Thus, the equation that we will implement takes the form:

$$\frac{\partial u}{\partial t} = v$$
$$\frac{\partial v}{\partial t} = c^2 \frac{\partial^2 u}{\partial x^2}. \tag{9.11}$$

This doubles the number of unknowns, as we now have two vectors to integrate ($u$ and $v$), both with length equal to the number of grid points.

### 9.2.2.2  Solving the Wave Equation in R

The implementation in R starts by defining the box sizes dx and the grid, xgrid. To comply with the boundary conditions (9.9) which are defined at $\pm\infty$, the grid needs to be taken large enough such that $u$ remains effectively 0 at the boundaries, for all times at which we calculate the solution.

Here, the grid extends from $-100$ to $100$; the number of grid cells (N) is returned by function setup.grid.1D. We will need it, as it defines the dimension of the system to solve (see below).

```
library(ReacTran)
dx      <- 0.2
xgrid <- setup.grid.1D(x.up = -100, x.down = 100, dx.1 = dx)
x       <- xgrid$x.mid
N       <- xgrid$N
```

The initial condition yini (9.8), and the output times are defined next:

```
lam    <- 0.05
uini   <- exp(-lam*x^2)
vini   <- rep(0, N)
yini   <- c(uini, vini)
times <- seq (from = 0, to = 50, by = 1)
```

The wave equation function (wave) first extracts, from the state variable vector y the two quantities u, v, both of length N, after which the **ReacTran** function tran.1D performs transport of u; the squared velocity ($c^2$) is taken as 1 (D = 1).

The function returns the derivatives of both u and v, combined (c ()) and packed as a list.

```
wave <- function (t, y, parms) {
    u <- y[1:N]
    v <- y[(N+1):(2*N)]

    du <- v
    dv <- tran.1D(C = u, C.up = 0, C.down = 0, D = 1,
```

```
                    dx = xgrid)$dC

    return(list(c(du, dv)))
}
```

In contrast to the previous (heat) problem, which was quite stiff and needed an implicit solver, this problem is not stiff at all, and can be solved most efficiently with the "adams" method. We also provide the dimension of the system and the names of the two quantitites. After the solution has been calculated, the property u is extracted from the output matrix out.

```
out <- ode.1D(func = wave, y = yini, times = times,
              parms = NULL, method = "adams",
              dimens = N, names = c("u", "v"))
u <- subset(out, which = "u")
```

The analytic solution (9.10) for each time-space point is now computed. R function outer, applies function analytic to each (times, x) pair.

```
analytic <- function (t, x)
    0.5 * (exp(-lam * (x+1*t)^2 ) +exp(-lam * (x-1*t)^2) )
OutAna <- outer(times, x, FUN = analytic)
```

The maximal absolute deviation with the computed numerical solution is printed.

```
max(abs(u - OutAna))
```

```
[1] 0.002188562
```

We now plot the results (Fig. 9.2); initial condition in black, the values for selected time points in darkgrey; a legend with times is written. It is simplest to do this plotting using **deSolve**'s function matplot.1D.

Note how the initial pulse is split in two smaller pulses, travelling in both directions, with velocity $c = 1$, and maximal value half the original value. At $t = 10$, the two pulses are centred around $-10$ and $+10$.

```
outtime <- seq(from = 0, to = 50, by = 10)
matplot.1D(out, which = "u", subset = time %in% outtime,
     grid = x, xlab = "x", ylab = "u", type = "l",
     lwd = 2, xlim = c(-50, 50),
     col = c("black", rep("darkgrey", 5)))
legend("topright", lty = 1:6, lwd = 2,
     col = c("black", rep("darkgrey", 5)),
     title = "t = ", legend = outtime)
```

You may also want to try the following "movie" (not shown):

```
plot.1D(out, grid = x, which = "u", type = "l",
     lwd = 2, ylim = c(0,1), ask = TRUE)
```

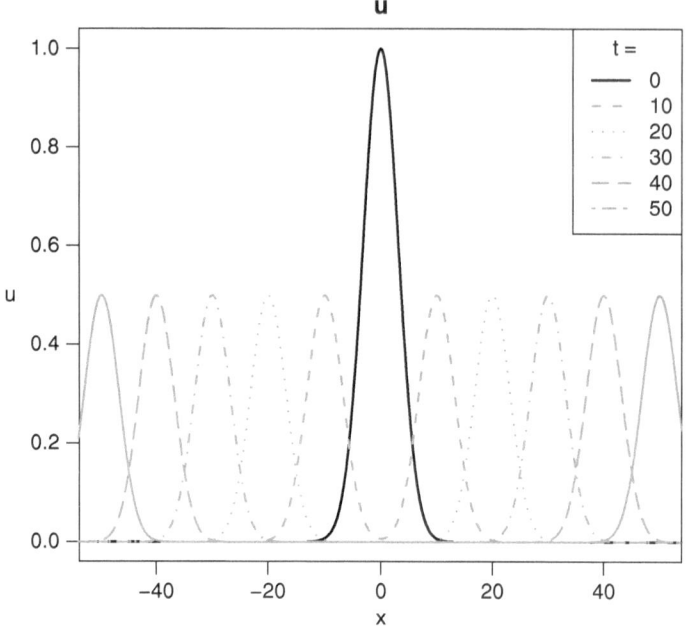

**Fig. 9.2** The 1-D wave equation; *black* = initial condition; *grey*: several time lines. See text for the R code

## 9.2.3   Poisson and Laplace's Equation

The *elliptic* partial differential equation, known as Poisson's equation is:

$$\nabla \cdot (\nabla w) = f, \tag{9.12}$$

where $f$ is a consumption rate.

In two dimensions, and for cartesian coordinates $(x, y)$, this becomes:

$$\frac{\partial^2 w}{\partial x^2} + \frac{\partial^2 w}{\partial y^2} = f(x, y). \tag{9.13}$$

### 9.2.3.1   Problem Definition

If $f(x, y) = 0$, (9.13) is called Laplace's equation.

With two spatial independent variables $(x, y)$, and no initial value variable, this second order equation requires two boundary conditions in the $x$ and two in the $y$ variable to be fully specified.

We define the equation in the $[0,1] \times [0,1]$ domain,[1] with boundaries:

$$
\begin{aligned}
w(x = 0, y) = w(x = 1, y) &= 0 \\
\frac{\partial w(x, y = 0)}{\partial y} &= 0 \\
\frac{\partial w(x, y = 1)}{\partial y} &= \sin(\pi x)\pi \sinh(\pi).
\end{aligned}
\tag{9.14}
$$

This has the following simple analytic solution:

$$
w(x, y) = \sin(\pi x)\cosh(\pi y).
\tag{9.15}
$$

### 9.2.3.2 Solving the Laplace Equation in R

Implementation in R follows the usual procedure of first defining a grid, in both directions (xgrid, ygrid). After that the model function (laplace) is implemented. Here the state variable vector U is first cast in matrix form w and then **ReacTran**'s appropriate transport function, tran.2D is called. It should be noted that none of the transport functions in **ReacTran** allows the specification of the boundaries as a *gradient*, but rather require input of the *flux*. With $flux = -D\partial w/\partial x$, it is however simple to convert one to the other. Here D.x and D.y are $= 1$:

```
Nx <- 100
Ny <- 100
xgrid <- setup.grid.1D (x.up = 0, x.down = 1, N = Nx)
ygrid <- setup.grid.1D (x.up = 0, x.down = 1, N = Ny)
x      <- xgrid$x.mid
y      <- ygrid$x.mid
```

```
laplace <- function(t, U, parms) {
    w <- matrix(nrow = Nx, ncol = Ny, data = U)
    dw <- tran.2D(C = w, C.x.up = 0, C.x.down = 0,
                  flux.y.up = 0,
                  flux.y.down = -1 * sin(pi*x)*pi*sinh(pi),
                  D.x = 1, D.y = 1,
                  dx = xgrid, dy = ygrid)$dC
    list(dw)
}
```

To solve the model, we cannot use the initial value solvers as in previous examples, as the equation is independent of time. Instead, we use root solver steady.2D from the R package **rootSolve** which is especially designed for solving 2-D equations. This solver requires input of the dimensions of the system (dimens),

---

[1] From http://www.scholarpedia.org/article/Partial_differential_equation/Approximate_and_Numerical_Methods.

a two-valued vector with the number of boxes in the $x$ and $y$-direction, the number of dependent variables (nspec) and the size of the work vector (lrw).[2] Another required input to steady.2D is an "initial guess" (y) of the solution. Fortunately, this (linear) set of equations is so simple to solve that any guess will do; here we just take Nx*Ny randomly distributed numbers (runif).

It takes less than a second to find the solution to these 10,000 equations:

```
print(system.time(
    out <- steady.2D(y = runif(Nx*Ny), func = laplace,
                    parms = NULL, nspec = 1,
                    dimens = c(Nx, Ny), lrw = 1e7)
))
```

```
 user   system elapsed
 0.37    0.02    0.39
```

We recast the solution in matrix form to compare it with the analytic solution (9.15):

```
w <- matrix(nrow = Nx, ncol = Ny, data = out$y)
analytic <- function (x, y) sin(pi*x) * cosh(pi*y)
OutAna <- outer(x, y, FUN = analytic)
max(abs(w - OutAna))
```

```
[1]  0.0006024049
```

We plot the results, using the image method for steady.2D output; we make the figure self-explanatory by adding contours (Fig. 9.3).

```
image(out, grid = list(x, y), main = "elliptic Laplace",
      add.contour = TRUE)
```

### 9.2.4   The Advection Equation

The advection equation is another hyperbolic equation that describes the transport of a quantity, $Y$, in a velocity field.

#### 9.2.4.1   Problem Definition

For the one-dimensional case, in a model domain $[a, b]$, we will solve the following advection equation:

$$\frac{\partial Y}{\partial t} = -v \frac{\partial Y}{\partial x}, \qquad (9.16)$$

---

[2]The latter requirement is unfortunate and it may require some trial and error to find a good value; however, if too little memory is allocated, the solver may stop with a message telling the size this vector should minimally have.

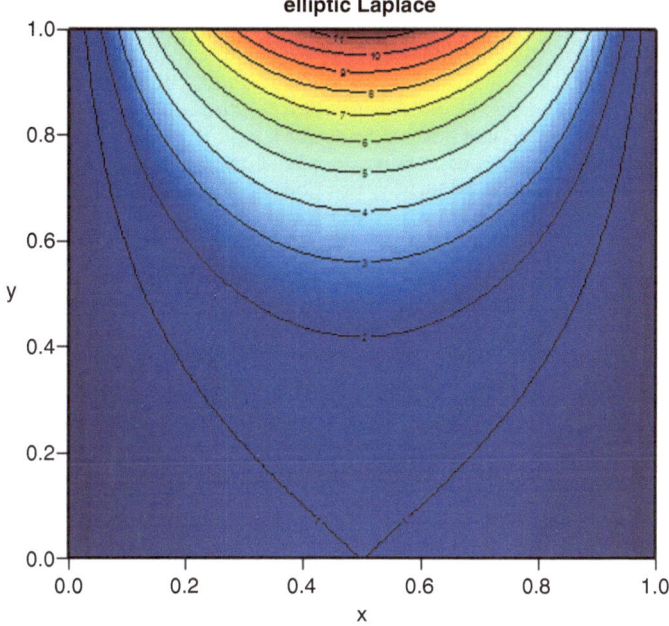

**Fig. 9.3** Solution of the 2-D laplace equation. See text for the R code

where $v$ is the constant advection rate $(= 0.1)$ and with initial and boundary conditions

$$Y(t = 0, x) = f$$
$$Y(t, x = a) = g. \tag{9.17}$$

The equation, which is first order in time and space needs only one initial value and boundary condition (for positive $v$, at $x = a$).

Despite its apparent simplicity, this equation is quite challenging to solve, especially if discontinuous, "shock" pulses are to be transported (see Sect. 8.3.4.1).

### 9.2.4.2   Solving the Advection Equation in R

We will not work out a complete new example, but just give the code that solved the model as in Fig. 8.4, where we tested several flux limiters. Here we solve the problem with flux limiters muscl and superbee (`"muscl"`, `"super"`).

```
adv.func <- function(t, y, p, adv.method)
    list(advection.1D(C = y, C.up = y[N], C.down = y[1],
                      v = 0.1, adv.method = adv.method,
                      dx = xgrid)$dC)
xgrid <- setup.grid.1D(0.3, 1.3, N = 50)
x       <- xgrid$x.mid
```

```
N       <- length(x)
yini    <- sin(pi * x)^50
times <- seq(0, 20, 0.01)
out1 <- ode.1D(y = yini, func = adv.func, times = times,
               parms = NULL, method = "euler", dimens = N,
               adv.method = "muscl")
out2 <- ode.1D(y = yini, func = adv.func, times = times,
               parms = NULL, method = "euler", dimens = N,
               adv.method = "super")
```

If you run this, you will notice that, even in the schemes that correct for numerical dispersion, there is still quite a lot of numerical artifact! Try:

```
plot.1D(out1, ylim = c(0, 1), type = "l", lwd = 2,
     main = "muscl")
plot.1D(out2, ylim = c(0, 1), type = "l", lwd = 2,
     main = "superbee")
```

## 9.3  More Complex Examples

We now solve some more complex equations, which are variations on the elliptic, parabolic and hyperbolic themes.

### 9.3.1  The Brusselator in One Dimension

Some partial differential equations can produce spatial patterns from an arbitrary initial state. These so-called Turing patterns [13] occur under certain conditions in coupled models of reacting and diffusing chemicals:

$$
\begin{aligned}
\frac{\partial U}{\partial t} &= D_U \nabla^2 U + f(U,V) \\
\frac{\partial V}{\partial t} &= D_V \nabla^2 V + g(U,V).
\end{aligned} \tag{9.18}
$$

The Brusselator was proposed [4] as a model for an auto-catalytic chemical reaction, between two products, $A$ and $B$, and producing also $C$ and $D$ in a number of intermediary steps. The chemical reactions are given by:

$$
\begin{aligned}
A &\xrightarrow{k_1} X_1 \\
B + X_1 &\xrightarrow{k_2} X_2 + C \\
2X_1 + X_2 &\xrightarrow{k_3} 3X_1 \\
X_1 &\xrightarrow{k_4} D,
\end{aligned} \tag{9.19}
$$

where the $k_i$ are the reaction rate constants. Assuming that the concentrations of $A$ and $B$ are kept constant, at values $a$ and $b$ respectively, the equations governing the dynamics of $X_1$ and $X_2$ are:

$$X_1' = k_1 a - k_2 b X_1 + k_3 X_1^2 X_2 - k_4 X_1$$
$$X_2' = k_2 b X_1 - k_3 X_1^2 X_2. \tag{9.20}$$

Putting $k_1, \ldots, k_4$, and $a = 1$ and $b = 3$ and adding diffusion we obtain

$$\frac{\partial X_1}{\partial t} = D_{X_1} \nabla^2 X_1 + 1 + X_1^2 X_2 - 4 X_1$$
$$\frac{\partial X_2}{\partial t} = D_{X_2} \nabla^2 X_2 + 3 X_1 - X_1^2 X_2. \tag{9.21}$$

It is instructive to implement the Brusselator first in one spatial dimension, and on a numerical grid composed of 50 boxes (N). We take for the boundary concentrations a value of 1 and 3 for $X_1$ and $X_2$ respectively. The diffusion coefficient, D, is taken to be 0.02.

```
library(ReacTran)
N      <- 50
Grid <- setup.grid.1D(x.up = 0, x.down = 1, N = N)
```

We choose interesting initial conditions:

```
x1ini <- 1 + sin(2 * pi * Grid$x.mid)
x2ini <- rep(x = 3, times = N)
yini <- c(x1ini, x2ini)
```

The derivative function is:

```
brusselator1D <- function(t, y, parms) {

    X1 <- y[1:N]
    X2 <- y[(N+1):(2*N)]

    dX1 <- 1 + X1^2*X2 - 4*X1 +
           tran.1D (C = X1, C.up = 1, C.down = 1,
                    D = 0.02, dx = Grid)$dC
    dX2 <- 3*X1 - X1^2*X2 +
           tran.1D (C = X2, C.up = 3, C.down = 3,
                    D = 0.02, dx = Grid)$dC

    list(c(dX1, dX2))
}
```

The equations are solved using function ode.1D, and output generated for $t = 0, 0.1, 0.2, \ldots, 10$. The time it takes to solve the model is printed. Note that we specify the number of boxes (dimens), as well as the number of chemical species (nspec). We also pass the names of the chemical species, which will facilitate plotting the output.

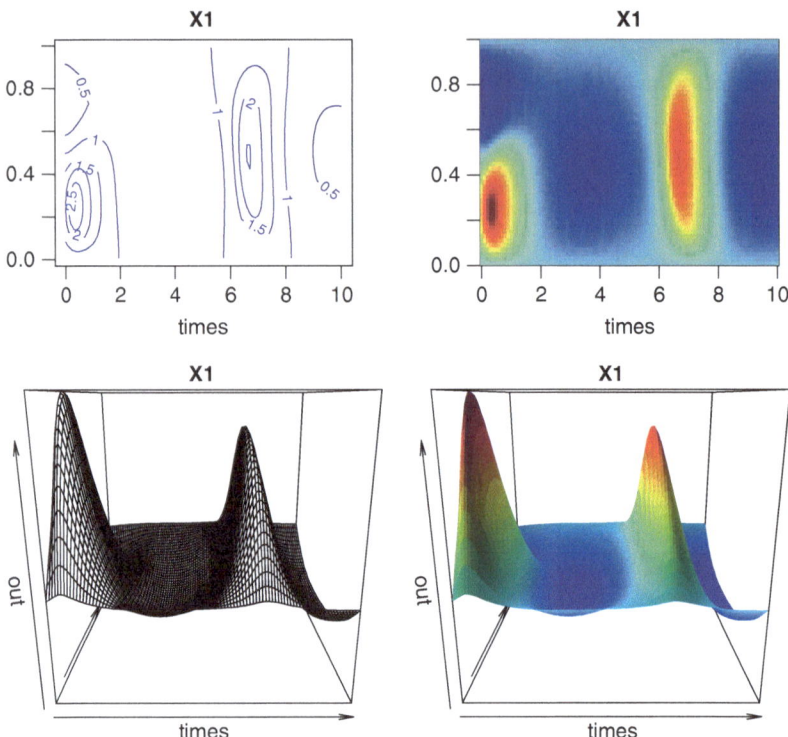

**Fig. 9.4**  Four different ways in which **deSolve**'s `image` plot can be used to depict the first variable of the 1-D Brusselator model. See text for the R code

```
times <- seq(from = 0, to = 10, by = 0.1)
print(system.time(
   out <- ode.1D(y = yini, func = brusselator1D,
                 times = times, parms = NULL, nspec = 2,
                 names = c("X1", "X2"), dimens = N)
))
```

```
 user   system elapsed
 0.33     0.00    0.34
```

We take the opportunity to show the various ways in which **deSolve**'s plotting method `image` can be used to display the output (Fig. 9.4). We start by specifying the number of figures in a row (`mfrow = c(2,2)`); the subsequent calls to `image` then pass `mfrow = NULL` to avoid the function overruling this property. The first variable (`which = "X1"`) is then plotted, first as a simple `contour` plot, then as a `filled.contour` (the default), and then twice as a `persp` plot; the first time without colour added (`col = NA`), the second time using the default colour scheme (which need not be specified), and adding a certain shade. We also pass the positions in the middle of each grid cell (Grid$x.mid). Before making the persp plots, the margin size is reduced (mar).

```
par(mfrow = c(2, 2))
image(out, mfrow = NULL, grid = Grid$x.mid,
      which = "X1",  method = "contour")
image(out, mfrow = NULL, grid = Grid$x.mid,
      which = "X1")
par(mar = c(1, 1, 1, 1))
image(out, mfrow = NULL, grid = Grid$x.mid,
      which = "X1",  method = "persp", col = NA)
image(out, mfrow = NULL, grid = Grid$x.mid,
      which = "X1",  method = "persp", border = NA,
      shade = 0.3 )
```

### 9.3.2   The Brusselator in Two Dimensions

In the presence of diffusion and when implemented in 2-D, this simple chemical model (9.18) can exhibit pattern-forming (so-called Turing) instabilities. Thus, the system, when initiated from a random distribution quickly generates spectacular oscillations or chaotic spatial concentration patterns.

The model in 2-D, implemented in R is very similar to the 1-D implementation:

```
brusselator2D <- function(t, y, parms) {

    X1 <- matrix(nrow = Nx, ncol = Ny,
                 data = y[1:(Nx*Ny)])
    X2 <- matrix(nrow = Nx, ncol = Ny,
                 data = y[(Nx*Ny+1) : (2*Nx*Ny)])

    dX1 <- 1 + X1^2*X2 - 4*X1 +
           tran.2D (C = X1, D.x = D_X1, D.y = D_X1,
                    dx = Gridx, dy = Gridy)$dC
    dX2 <- 3*X1 - X1^2*X2 +
           tran.2D (C = X2, D.x = D_X2, D.y = D_X2,
                    dx = Gridx, dy = Gridy)$dC

    list(c(dX1, dX2))
}
```

Note that we have imposed zero-gradient boundaries in the $x-$ and $y-$ direction. As this is the default, the boundary conditions need not be explicitly specified.

The numerical grid is composed of 50 boxes in the $x-$ and $y-$ directions, extending from 0 to 1.

```
library(ReacTran)
Nx    <- 50
Ny    <- 50
Gridx <- setup.grid.1D(x.up = 0, x.down = 1, N = Nx)
Gridy <- setup.grid.1D(x.up = 0, x.down = 1, N = Ny)
```

We choose parameter values that give interesting patterns (these require at least the two diffusion coefficients to be different):

```
D_X1 <- 2
D_X2 <- 8*D_X1
```

As initial condition for the two chemical substances, simple random numbers in between 0 and 1 are used:

```
X1ini <- matrix(nrow = Nx, ncol = Ny, data = runif(Nx*Ny))
X2ini <- matrix(nrow = Nx, ncol = Ny, data = runif(Nx*Ny))
yini <- c(X1ini, X2ini)
```

The model is solved using **deSolve** function ode.2D and output generated for t = 0, 1, . . . 8. The size of the work space (lrw), the number of modeled components (nspec) and their names, and the dimensions of the domain (dimens) is given.

```
times <- 0:8
print(system.time(
   out <- ode.2D(y = yini, parms = NULL, func = brusselator2D,
                 nspec = 2, dimens = c(Nx, Ny), times = times,
                 lrw = 2000000, names=c("X1", "X2"))
))
```

```
   user   system elapsed
   2.78    0.00    2.83
```

Finally the output is plotted (Fig. 9.5) using **deSolve**'s function image. For the main title above each figure (main) we write the time; a global title is written with mtext; we first increase the size of the outer margin (oma).

```
par(oma = c(0,0,1,0))
image(out, which = "X1", xlab = "x", ylab = "y",
      mfrow = c(3, 3), ask = FALSE,
      main = paste("t = ", times),
      grid = list(x = Gridx$x.mid, y = Gridy$x.mid))
mtext(side = 3, outer = TRUE, cex = 1.25, line = -1,
      "2-D Brusselator, species X1")
```

Already at $t = 1$, the initial random pattern has given way to a quite structured spatial pattern. With these parameter values, usually a random pattern returns at around $t = 8$ (but every run is different).

### 9.3.3  Laplace Equation in Polar Coordinates

We now implement the Laplace equation defined on an annulus-shaped domain extending from $r = 2$ to $r = 4$, and with prescribed values on the two boundaries.[3]

---

[3]From http://en.wikipedia.org/wiki/Laplace's_equation.

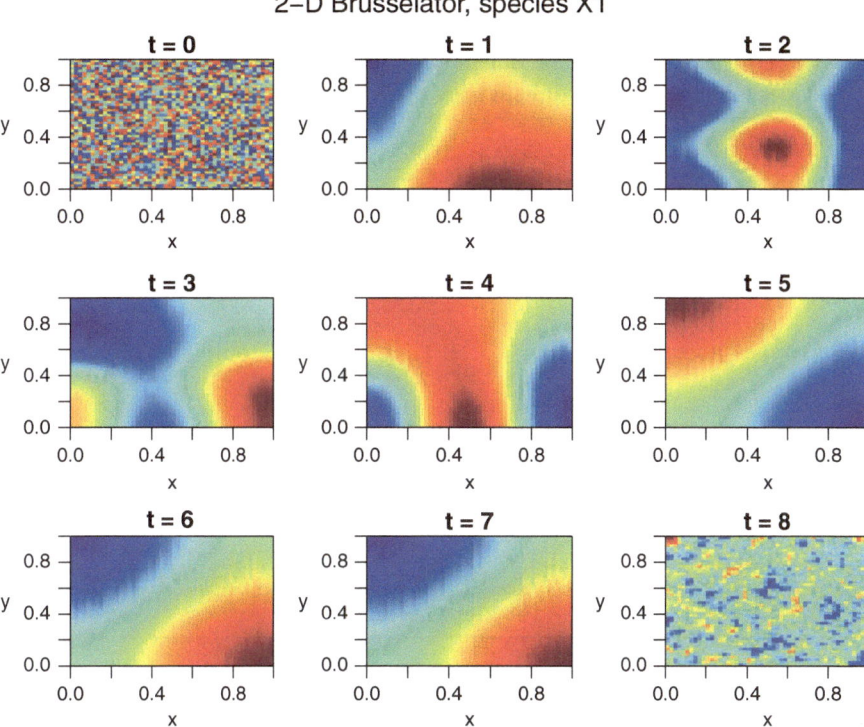

**Fig. 9.5** Solution of the 2-D Brusselator, species X1. See text for the R code

As the model domain is circular, this model is best described in polar coordinates (see Sect. 8.1.2):

$$\frac{\partial u}{\partial t} = \frac{1}{r}\frac{\partial}{\partial r}\left(r\frac{\partial u}{\partial r}\right) + \frac{1}{r^2}\frac{\partial^2 u}{\partial \theta^2}. \tag{9.22}$$

The boundaries are set to a constant value (0) in the centre of the domain at $r = 2$, while at the outer edge it is prescribed by a sine wave, as a function of $\theta$.

$$\begin{aligned} u(t, r = 2, \theta) &= 0 \\ u(t, r = 4, \theta) &= 4\sin(5\theta). \end{aligned} \tag{9.23}$$

We also need boundary conditions in the direction of $\theta$, which extends from 0 to $2\pi$. As the two edges connect in the $\theta$ direction (it is an annulus), we prescribe that the value and gradients at $\theta = 0$ and $\theta = 2\pi$ are the same, i.e. the boundary in the second dimension is cyclic.

$$\begin{aligned} u(t, r, \theta = 0) &= u(t, r, \theta = 2\pi) \\ \frac{\partial u}{\partial \theta}(t, r, \theta = 0) &= \frac{\partial u}{\partial \theta}(t, r, \theta = 2\pi). \end{aligned} \tag{9.24}$$

**ReacTran** function `tran.polar` performs transport in polar coordinates. The input slightly differs from the other transport functions, in that here the *positions on the interfaces* in `r` and $\theta$ direction must be input (arguments `r` and `theta`). The boundaries in the second ($\theta$) direction are cyclic (`cyclicBnd=2`).

```
library(ReacTran)
Nr <- 100
Np <- 100
r          <- seq(2, 4, len = Nr+1)
theta      <- seq(0, 2*pi, len = Np+1)
theta.mid <- 0.5*(theta[-1] + theta[-Np])
Model <- function(t, C, p) {
  y = matrix(nrow = Nr, ncol = Np, data = C)
  tran <- tran.polar (y, D.r = 1, r = r, theta = theta,
              C.r.up = 0, C.r.down = 4 * sin(5*theta.mid),
              cyclicBnd = 2)
  list(tran$dC)
}
```

The model is solved to steady-state with function `steady.2D`:

```
STD <- steady.2D(y = runif(Nr*Np), parms = NULL,
                 func = Model, dimens = c(Nr, Np),
                 lrw = 1e6, cyclicBnd = 2)
```

Before we plot the ouput using function `image` (Fig. 9.6), we map the polar (`r`, `theta`) to cartesian (`x`, `y`) coordinates.

```
OUT <- polar2cart (STD, r = r, theta = theta,
                   x = seq(-4, 4, len = 400),
                   y = seq(-4, 4, len = 400))
image(OUT, main = "Laplace")
```

### 9.3.4   The Time-Dependent 2-D Sine-Gordon Equation

The Sine-Gordon equation is a non-linear hyperbolic (wave-like) partial differential equation involving the sine of the dependent variable. The equation in two dimensions, defined on $[-7,7]$, is:

$$\frac{\partial^2 u}{\partial t^2} = D\frac{\partial^2 u}{\partial x^2} + D\frac{\partial^2 u}{\partial y^2} - \sin u, \tag{9.25}$$

with boundary and initial conditions equal to:

$$
\begin{aligned}
u(t,x=-7,y) &= u(t,x=7,y) = u(t,x,y=-7) = u(t,x,y=7) = 0 \\
u'(t=0,x,y) &= 0 \\
u(t=0,x,y) &= exp(-((x-2)^2+(y-2)^2))+exp(-((x-2)^2+(y+2)^2))+ \\
&\quad exp(-((x+2)^2+(y-2)^2))+exp(-((x+2)^2+(y+2)^2)).
\end{aligned}
$$

$$\tag{9.26}$$

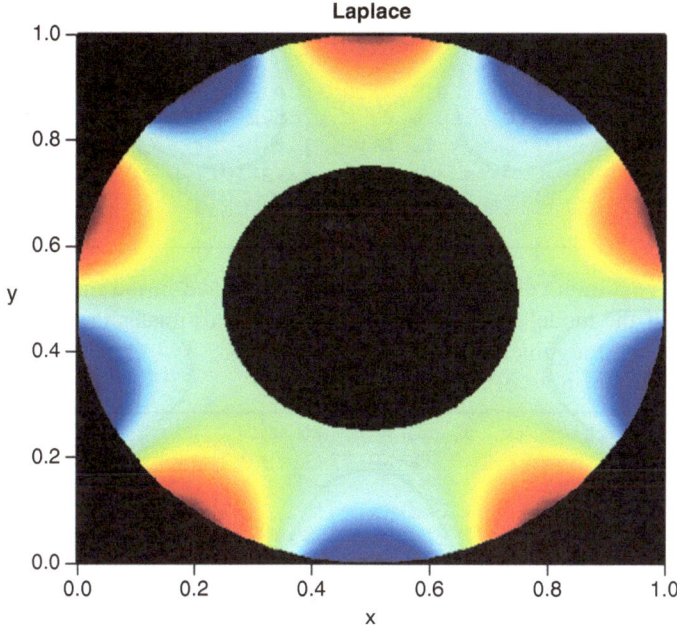

**Fig. 9.6** Solution of the Laplace equation, in polar coordinates. See text for the R code

Similarly to what we did for the wave equation (Sect. 9.2.2), this model is rewritten as two first order differential equations.

$$\frac{du}{dt} = v$$
$$\frac{\partial v}{\partial t} = D\frac{\partial^2 u}{\partial x^2} + D\frac{\partial^2 u}{\partial y^2} - \sin u. \tag{9.27}$$

The model domain $[-7,7]$ is divided into 100 boxes in both directions. As there are two dependent variables $(u,v)$, this leads to a system comprising 20,000 equations.

```
Nx    <- 100
Ny   <- 100
xgrid <- setup.grid.1D(-7, 7, N=Nx)
ygrid <- setup.grid.1D(-7, 7, N=Ny)
x <- xgrid$x.mid
y <- ygrid$x.mid
```

The problem has prescribed values $(= 0)$ at the four boundaries (C.x.up, C.x.down, ...). **ReacTran** function tran.2D performs the diffusive transport.

```
sinegordon2D <- function(t, C, parms) {

    u <- matrix(nrow = Nx, ncol = Ny,
             data = C[1 : (Nx*Ny)])
```

```
      v <- matrix(nrow = Nx, ncol = Ny,
              data = C[(Nx*Ny+1) : (2*Nx*Ny)])

      dv <- tran.2D (C = u, C.x.up = 0, C.x.down = 0,
                     C.y.up = 0, C.y.down = 0,
                     D.x = 1, D.y = 1,
                     dx = xgrid, dy = ygrid)$dC - sin(u)
      list(c(v, dv))
    }
```

The initial condition consists of four peaks, positioned in the middle of the four quadrants of the model domain. A function, peak, estimates the peak values for one set of (x, y) arguments. R function outer (x, y, FUN = ...) applies FUN to all combinations of x and y.

```
  peak <- function (x, y, x0 = 0, y0 = 0)
                  exp(-((x-x0)^2 + (y-y0)^2))
  uini <- outer(x, y,
    FUN = function(x, y) peak(x, y, 2,2) + peak(x, y,-2,-2)
                       + peak(x, y,-2,2) + peak(x, y, 2,-2))
  vini <- rep(0, Nx*Ny)
```

We produce output only at four time values. As this problem is non-stiff, we use a fifth order Runge-Kutta scheme to perform the integration (method="ode45", also known as DOPRI5):

```
  times <- 0:3
  print(system.time(
    out   <- ode.2D (y = c(uini, vini), times = times,
                   parms = NULL, func = sinegordon2D,
                   names = c("u", "v"),
                   dimens = c(Nx, Ny), method = "ode45")
  ))
```

```
    user   system elapsed
    0.75    0.00    0.77
```

Before plotting the output, as an image (Fig. 9.7), the margin size is enlarged (par (mar))

```
  mr <- par(mar = c(0, 0, 1, 0))
  image(out, main = paste("time =", times), which = "u",
        grid = list(x = x, y = y), method = "persp",
        border = NA, col = "grey",  box = FALSE,
        shade = 0.5, theta = 30, phi = 60, mfrow = c(2, 2),
        ask = FALSE)
  par(mar = mr)
```

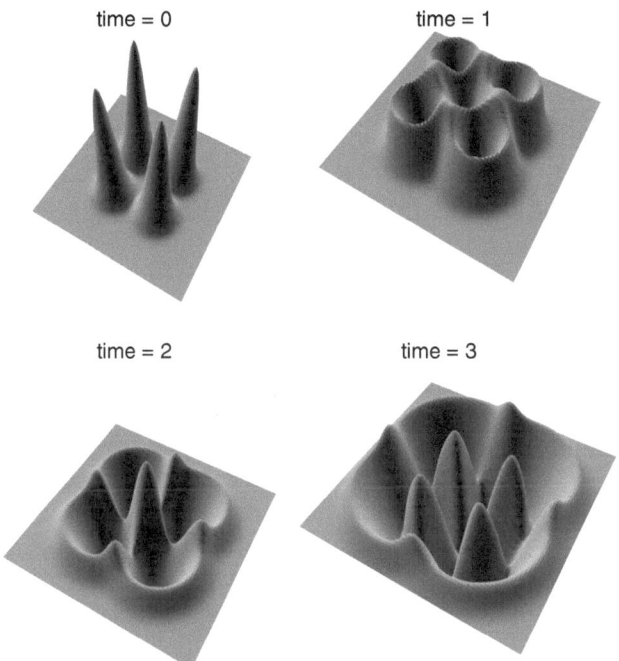

**Fig. 9.7** Solution of the 2-D sine-gordon equation. See text for the R code

The following code gives a more `movie-like` output (not shown):

```
out  <- ode.2D (y = c(uini, vini), times = seq(0, 3, by = 0.1),
            parms = NULL, func = sinegordon2D,
            names=c("u", "v"), dimens = c(Nx, Ny),
            method = "ode45")
image(out, which = "u", grid = list(x = x, y = y),
    method = "persp", border = NA,
    theta = 30, phi = 60, box = FALSE, ask = FALSE)
```

### 9.3.5 The Nonlinear Schrödinger Equation

The Schrödinger equation is one of the fundamental equations of quantum mechanics.

We implement a nonlinear equation [8] used to model for instance solitons in optical fibre pulse propagation. It is given by:

$$\frac{\partial u}{\partial t} = i\frac{\partial^2 u}{\partial x^2} + i\gamma|u|^2 u, \tag{9.28}$$

where $u$ is complex, $\gamma$ is a real constant, and $i = \sqrt{-1}$.

As for any diffusion (parabolic) equation, the Schrödinger equation is solved by temporal integration of spatial discretisations, as generated by function `tran.1D` from the package **ReacTran**:

```
alf <- 0.5
gam <- 1
Schrodinger <- function(t, u, parms) {
    du <- 1i * tran.1D (C = u, D = 1, dx = xgrid)$dC  +
                  1i * gam * abs(u)^2 * u
    list(du)
}
```

where `1i` is $\sqrt{-1}$ and where we have assumed zero-gradient boundary conditions. As these are the default boundary conditions they need not be specified.

To solve the model in a qualitatively correct way, the model domain must be divided into a sufficient number of boxes [2]; here we take $N = 300$.

```
N        <- 300
xgrid <- setup.grid.1D(-20,  80,  N = N)
x        <- xgrid$x.mid
```

For a single soliton, the above Schrödinger equation has as solution:

$$u(t,x) = \sqrt{\frac{2\alpha}{\gamma}} \exp(i(0.5cx - t(1/4c^2 - \alpha)))sech(\sqrt{\alpha}(x - ct)), \qquad (9.29)$$

where $c$ is the speed at which it travels, and where $\alpha$ determines the soliton's amplitude, and $sech(x) = 2/(e^x + e^{-x})$.

To demonstrate the peculiarities of solitons, we model *two* of them, a fast moving one (velocity $c_1$) and a slower one ($c_2$). Initially, both are well separated, the fast one located at $x = 0$, the slowly moving soliton centred at $x = 25$.

The initial condition for each of these solitons, can be easily derived from the above analytic solution by setting $t = 0$, and using $c_1$ and $c_2$ for the two velocities.

To calculate the initial conditions in R, it is easiest to define two functions, one that returns $sech(x)$ and a second function that estimates the profile for one soliton, using (9.29); the initial condition `yini` is then the sum of the two solitons.

```
c1    <- 1
c2    <- 0.1
sech      <- function(x) 2/(exp(x) + exp(-x))
soliton <- function (x, c1)
    sqrt(2*alf/gam)  * exp(0.5*1i*c1*x)  * sech(sqrt(alf)*x)
yini <- soliton(x, c1)  + soliton(x-25,  c2)
```

It is fastest to solve the model with the "adams" method:

```
times <- seq(0,  40,  by = 0.1)
print(system.time(
  out <- ode.1D(y = yini, parms = NULL, func = Schrodinger,
```

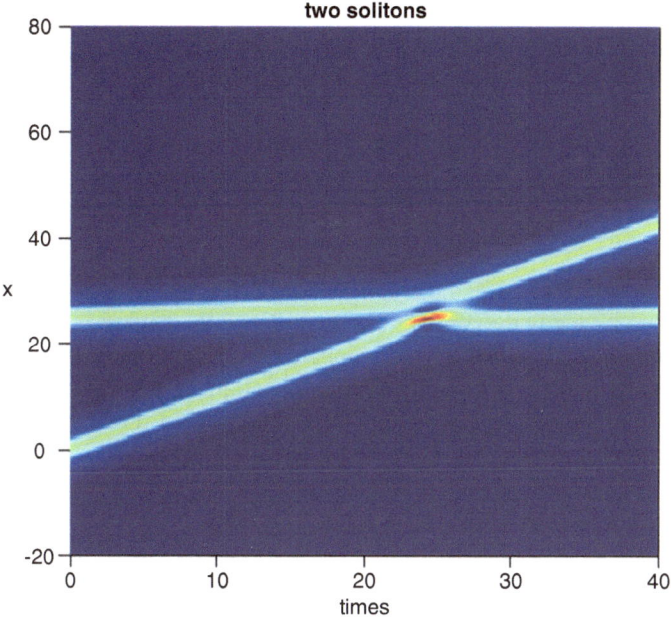

**Fig. 9.8** Solution of the Schrödinger equation. See text for the R code

```
                    times = times, dimens = 300, method = "adams")
))
```

```
user  system elapsed
2.61    0.01    2.65
```

The model output (Fig. 9.8) shows how both solitons travel through the model domain; as the faster wave collides with the slower one, it passes through it, and the shape and velocities of both solitons remain the same afterwards.

```
image(abs(out), grid = x, ylab = "x", main = "two solitons")
```

## 9.4 Exercises

### 9.4.1 The Gray-Scott Equation

Interesting patterns are also generated with the Gray-Scott equation [6].
It describes the following chemical reaction:

$$U + 2V \xrightarrow{k_0} 3V$$
$$V \xrightarrow{k_1} P.$$

(9.30)

The two species disperse on a rectangular grid, and both are diluted at rate $f$; U is added at the same rate, and with input concentration $= U_{in}$.

The mathematical model is:

$$\frac{\partial U}{\partial t} = D_U \frac{\partial^2 U}{\partial x^2} - k_0 UV^2 - fU + fU_{in}$$
$$\frac{\partial V}{\partial t} = D_V \frac{\partial^2 V}{\partial x^2} + k_0 UV^2 - fV - k_1 V. \qquad (9.31)$$

The equation is described on the grid $0 \leq x, y \leq 2.5$

Similarly to what is the case for the Brusselator, the pattern generated depends on the subtle differences in the rates, compared to the diffusion coefficients.

For the parameters, use $D_U = 8e{-}5$ and $D_V = 0.5 D_U$; $f = 0.024$, $U_{in} = 1$, $k_0 = 1$ and $k_1 = 0.06$. Compose a grid of 100 boxes in the $x$ and $y$ direction.

Use as initial conditions the following pattern:

$$U(t = 0, x, y) = 1 - 2V(t = 0, x, y)$$
$$V(t = 0, x, y) = 0.25 \sin^2(4\pi x) \sin^2(4\pi y) \text{ for } 1 \leq x, y \leq 1.5 \qquad (9.32)$$
$$V(t = 0, x, y) = 0 \qquad\qquad\qquad\text{elsewhere.}$$

Try to create Fig. 9.9:

### 9.4.2  A Macroscopic Model of Traffic

Estimation of travel times on highways can be performed by application of a macroscopic model, where traffic is described in an analogous way to fluid flow [3]. Conservation of cars on a highway, in the domain $[0, L]$ leads to the equation:

$$\frac{\partial \rho(t, x)}{\partial t} = -\frac{\partial v(t, x) \rho(t, x)}{\partial x}, \qquad (9.33)$$

where $\rho(t, x)$ is the density of traffic, as a function of position $x$ and time $t$, and where $v(t, x)$ is the velocity, given by :

$$v = v_f \left(1 - \frac{\rho}{\rho_{max}}\right) - D\frac{\partial \rho}{\partial x}\Big/\rho, \qquad (9.34)$$

where $v_f$ is the free-flow speed, $\rho_{max}$ is the car density in a traffic jam, and $D$ is a "diffusion coefficient" that takes into account the fact that drivers can adapt their speed depending on the density of cars ahead of them. The first part of the equation ensures that cars move slower if the traffic density is higher, and will stop in a traffic jam. The second "diffusive" part can also be understood to mimic the variable reaction speed of car drivers, e.g. after traffic light switches.

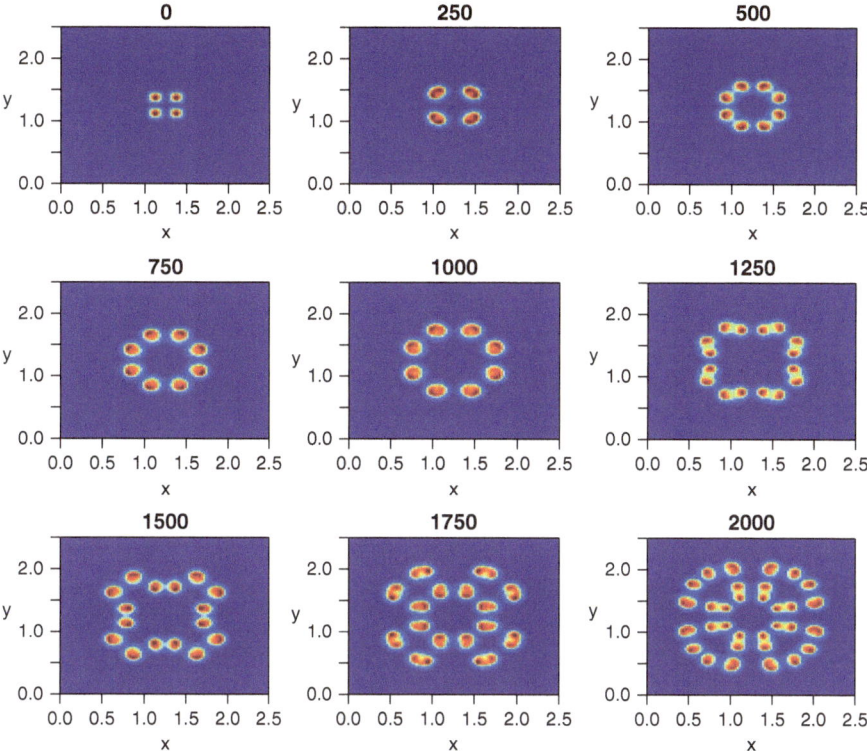

**Fig. 9.9** The Gray-Scott equation

Assume a red light at position $x_0 = 10$ (m), the initial condition becomes $\rho(0,x) = \rho_{max}$ for $x < x_0$ and 0 for $x > x_0$.

Implement how a line of cars moves along a Belgian road where the speed limit in an urban area is 50 km/h; assume all cars try to drive at that speed.

### 9.4.3 A Vibrating String

On my cello, both ends of the four strings are fixed, one end at the bridge, one at the nut (headstock). The length of the string is set by the distance between these two fixed ends. Defining the coordinate system such that one end of the string corresponds with the bridge at $x = 0$, the other end with the nut, at $x = L$, the boundary conditions for the string wave are such that:

$$
\begin{aligned}
u(t,0) &= 0 \\
u(t,L) &= 0,
\end{aligned}
\tag{9.35}
$$

for all times.

Assume that the instrument is played in pizzicato; model a wave in an ideal string (not a cello). Take $L = 1$. For initial condition of the position $u$, and velocity $u'$ take:

$$u(t = 0, x) = \exp(-\lambda (x - 0.2)^2)$$
$$u'(t = 0, x) = 0. \tag{9.36}$$

Note that, after plucking a string, two waves are set in motion, moving in opposite directions; when reaching the ends, they reflect. Use plot.1D to see how the wave travels across the string.

### 9.4.4  A Pebble in a Bucket of Water

Imagine throwing a pebble into the middle of a bucket of water. As it touches the surface, waves are generated that propagate in all directions. The waves retain their shape until they hit the wall of the bucket where they reflect.

As the bucket is circular, this model is best solved using polar coordinates. Moreover, as the pebble is dropped in the centre, we can represent the wave propagation by a 1-D model:

$$\frac{\partial^2 u}{\partial t^2} = c^2 \frac{1}{r} \frac{\partial}{\partial r} \left( r \frac{\partial u}{\partial r} \right)$$
$$u_{t=0,r} = -\exp(-0.2 r^2)$$
$$\frac{\partial u}{\partial r}_{t,0} = 0 \tag{9.37}$$
$$u_{t,a} = 0.$$

Implement this model in R , using functions tran.1D and ode.1D. Hint: use argument A for specifying $r$ in tran.1D. A similar example, of the parabolic type has been solved in **ReacTran**'s vignette called "PDE" (vignette("PDE")

### 9.4.5  Combustion in 2-D

A relatively stiff PDE is the combustion problem from [2]. This problem describes diffusion and reaction

$$\frac{\partial U}{\partial t} = -\nabla \cdot (-K\nabla U) + \frac{R}{\alpha \delta}(1 + \alpha - U)\exp(\delta(1 - 1/U)). \tag{9.38}$$

Implement this model on a rectangular 2-dimensional domain ($[0, 1] \times [0, 1]$). Use as values for the parameters $K = 1, \alpha = 1, \delta = 20, R = 5$, and as initial conditions use a constant: $U(0, x, y) = 1$. The behavior of the solution at the boundaries is prescribed as a known value (=1) for the downstream boundary, and a zero-flux boundary upstream. Integrate the model in $[0.0, 36]$ using `ode.2D`, and estimate the steady-state condition; the latter you can calculate by using **rootSolve**'s function `steady.2D`.

# References

1. Burchard, H., Bolding, K., & Villarreal, M. R. (1999). **GOTM**, *a general ocean turbulence model. Theory, applications and test cases* (Tech Rep EUR 18745 EN). European Commission.
2. Hundsdorfer, W., & Verwer, J. G. (2003). *Numerical solution of time-dependent advection-diffusion-reaction equations. Springer series in computational mathematics.* Berlin: Springer.
3. Kachroo, P., Ozbay, K., & Hobeika, A. G. (2001). Real-time travel time estimation using macroscopic traffic flow models. In *2001 IEEE intelligent transportation systems conference proceedings*, Oakland (pp. 132–137).
4. Lefever, R., Nicolis, G., & Prigogine, I. (1967). On the occurrence of oscillations around the steady state in systems of chemical reactions far from equilibrium. *Journal of Chemical Physics, 47*, 1045–1047.
5. Leonard, B. P. (1988). Simple high accuracy resolution programs for convective modeling of discontinuities. *International Journal for Numerical Methods in Fluids, 8*, 1291–1318.
6. Pearson, J. (1993). Complex patterns in a simple system. *Science, 261*, 189–192.
7. Roe, P. L. (1985). Some contributions to the modeling of discontinuous flows. *Lectures Notes in Applied Mathematics, 22*, 163–193. Amer. Math. Soc., Providence.
8. Sanz-Serna, J. M. (1984). Methods for the numerical solution of the nonlinear Schrodinger equation. *Mathematics of Computation, 43*, 21–27.
9. Soetaert, K. (2011). **rootSolve**: *Nonlinear root finding, equilibrium and steady-state analysis of ordinary differential equations.* R package version 1.6.2.
10. Soetaert, K., & Meysman, F. (2011). **ReacTran**: *Reactive transport modelling in 1D, 2D and 3D.* R package version 1.3.2.
11. Soetaert, K., & Meysman, F. (2012). Reactive transport in aquatic ecosystems: Rapid model prototyping in the open source software R. *Environmental Modelling and Software, 32*, 49–60.
12. Soetaert, K., Petzoldt, T., & Setzer, R. W. (2010). Solving differential equations in R: Package **deSolve**. *Journal of Statistical Software, 33*(9), 1–25.
13. Turing, A. M. (1952). The chemical basis of morphogenesis. *Philosophical Transactions of the Royal Society of London Series B, 237*, 37–72.
14. van Leer, B. (1979). Towards the ultimate conservative difference scheme V. A second order sequel to Godunov's method. *Journal of Computational Physics, 32*, 101–136.

# Chapter 10
# Boundary Value Problems

**Abstract** When solving initial value problems for ordinary differential equations, differential algebraic equations or partial differential equations, as discussed in previous chapters, a unique solution to the equations, if it exists, is obtained by specifying the values of all the components at the starting point of the range of integration. With boundary value problems (BVPs), the conditions are specified at more than one point, usually (but not necessarily) at the boundaries of the independent variable. Because of this it is not guaranteed that BVPs have a unique solution; they may have no solution at all or many solutions. The theory of BVPs, such as the proof of existence and uniqueness of solutions, is considerably more difficult than it is in the initial value case. Also, software for BVPs is much less well developed than for IVPs. In this chapter we will deal mainly with two-point boundary value problems which have the boundary conditions specified at both ends of a finite range of integration. We discuss two distinct methods to solve BVPs, namely shooting and finite difference methods.

## 10.1 Two-Point Boundary Value Problems

We define the general form taken by a two-point boundary value problem as:

$$y' = f(x,y), \quad a \le x \le b$$
$$g(y(a),y(b)) = 0, \tag{10.1}$$

where $y \in \Re^m$, $m > 1$. We recall that, as explained in Sect. 1.1.3, any high order differential equation can be converted into a system of first order equations. Therefore in the following we consider both first order and high order equations.

A nice example of a BVP of practical interest is the vibrating spring problem which is defined as [1, Chap. 1]:

$$-(p(x)y')' + q(x)y = r(x), \tag{10.2}$$

K. Soetaert et al., *Solving Differential Equations in R*, Use R!,
DOI 10.1007/978-3-642-28070-2_10, © Springer-Verlag Berlin Heidelberg 2012

where $p(x) > 0$ and $q(x) \geq 0$ for all $a \leq x \leq b$. If the spring is clamped at one end ($a$) and left free to oscillate at the other end ($b$), then we have, associated with (10.2), the boundary conditions

$$
\begin{aligned}
y(a) &= 0 \\
y'(b) &= 0.
\end{aligned}
\tag{10.3}
$$

Notice that these boundary conditions have a special form which is much less general than those appearing in (10.1). The boundary conditions in (10.3) are referred to as being *separated*, and we will return to this very important special case later. For a discussion of vibration problems the reader is referred to [38].

## 10.2   Characteristics of Boundary Value Problems

It is not our intention to go too deeply into the theory of boundary value problems (see [1] or [5] for that), but we will explain a few important aspects of the theory which we feel are directly relevant to our discussion.

### 10.2.1   Uniqueness of Solutions

Let us start with the IVP

$$
\begin{aligned}
y' &= f(x,y), \quad x \geq a \\
y(a) &= c,
\end{aligned}
\tag{10.4}
$$

where $y \in \Re^m$. We denote the solution of (10.4) by $y(x;c)$. The task of solving the BVP given by (10.1) now boils down to finding a solution such that the following holds:

$$
g(c;y(b;c)) = 0.
\tag{10.5}
$$

This gives a set of nonlinear algebraic equations for the unknown vector $c$. The number of solutions of problem (10.1) is equal to the number of solutions of the algebraic equations given by (10.5) and this number can be 0, 1 or many, (see [5], p. 89).

A famous example is the Bratu problem [18] which describes spontaneous combustion:

$$
\begin{aligned}
y'' + \lambda \exp(y) &= 0 \\
y(0) = y(1) &= 0.
\end{aligned}
\tag{10.6}
$$

For values of $\lambda$ smaller than $3.51383\ldots$, this equation has two solutions; above this value there is no solution at all; at the critical value of $\lambda$, there is exactly one solution [18].

### 10.2.2 Isolation of Solutions

If a given problem has, for example, two solutions then it may be possible to find both of them. We can often find multiple solutions by trying different initial guesses to the solution. The important question is then whether or not the solutions are *isolated*. If a solution is isolated, then there is no other solution of the problem in the neighborhood of this solution. Intuitively this means that we can put a tube around the solution and no other solution of our problem will venture into this tube. To simplify the situation, without loss of generality, we will consider the problem defined in the form:

$$y' = f(x,y), \quad a \leq x \leq b$$
$$B_a y(a) + B_b y(b) = rhs. \tag{10.7}$$

Here $B_a$ and $B_b \in \Re^{m \times m}$ and $rhs \in \Re^m$. The way we determine whether or not a solution $y(x)$ is isolated is to look at the *variational problem*. This problem is derived from (10.1) by perturbing the variables by small amounts and ignoring second order terms. This gives the linear problem:

$$Z' = J(x, y(x))Z, \quad a \leq x \leq b$$
$$B_a Z(a) + B_b Z(b) = 0. \tag{10.8}$$

Here $J$ is the Jacobian matrix $\partial f(x, y(x))/\partial y$. If this variational equation has only the unique solution $Z = 0$ then $y(x)$ is an isolated solution.

A special case of (10.5) is when the problem is linear as it is normally straightforward to determine how many solutions there are. In this case we can write down a condition in terms of the Green's function, which guarantees that the problem has a unique solution.

### 10.2.3 Stiffness of Boundary Value Problems and Dichotomy

A stiff problem is one which has some very fast modes in addition to some slow ones. A stiff *initial value problem* will be stable only if these rapid modes decrease with increasing time. One major difference between the concept of stiffness for IVPs and BVPs is that for BVPs we must allow both rapidly increasing and rapidly decreasing modes to be present in the solution.

Suppose that the solution of (10.1) has $k$ non-increasing modes and $m - k$ non-decreasing modes. It is important to define the BVP (10.1) so that the $k$ non-increasing modes are controlled by the boundary conditions at $x = a$ and the $m - k$ non-decreasing modes are controlled by the boundary conditions at $x = b$ (recall $a < b$). This concept is called dichotomy [5] and ensures the separation of modes throughout the interval. It is important because it is a necessary and sufficient condition for a BVP to be stable [5, p. 115].

### 10.2.4  Conditioning of Boundary Value Problems

As explained in Sect. 1.2.4 the concept of stability, which is usually applied to initial value problems for differential equations, is referred to as conditioning for boundary value problems. Conditioning relates to the effect small changes in (10.1), either in the function $f$, or in the boundary conditions $g(y(a), y(b))$, have on the solution.

This concept is discussed in detail in [5, Chap. 3, Chap. 5] where it is shown that, referring to the Jacobian matrix $M$ of the discretized problems, a solution method is well-conditioned if there exists a *stability (or conditioning) constant* $\kappa$ of moderate size such that $||M^{-1}|| \leq \kappa$.

An important remark on conditioning is made in [1, p. 203]: "if the BVP is stable and not stiff, and the local truncation error is small then the global error is expected to have the order of the local truncation error times the stability constant of the given BVP". This means that it will be very important, perhaps vital in some cases to obtain an estimate of $\kappa$ when computing a solution of a BVP, since a small local error does not necessarily give rise to a small global error.

In order to obtain a more precise feeling for the concept of a stable boundary value problem, we refer the reader to [1, p. 169]. Here stability is defined to mean that the conditioning constant associated with the differential equation is of moderate size. This in turn means that the Green's function is nicely bounded.

To introduce the conditioning parameters let us consider for simplicity the following linear boundary value problem:

$$\frac{dy}{dx} = A(x)y(x) + q(x), \ a \leq x \leq b, \ B_a y(a) + B_b y(b) = \beta, \beta \in R^m, \qquad (10.9)$$

whose solution is given by

$$y(x) = Y(x)Q^{-1}\beta + \int_a^b G(x, t)q(t)dt. \qquad (10.10)$$

Here $Y(x)$ is a fundamental solution, $Q = B_a Y(a) + B_b Y(b)$ is non singular and $G(x, t)$ is the Green's function. Using the $\infty$-norm we can compute the conditioning parameter by considering a perturbed equation:

$$\frac{du}{dx} = A(x)u(x) + q(x) + \delta(x), \ a \leq x \leq b, \ B_a u(a) + B_b u(b) = \beta + \delta\beta. \quad (10.11)$$

Here $\delta(x)$ and $\delta\beta$ are small perturbations of the data. The difference between the two solutions satisfies:

$$||u(x) - y(x)|| \leq ||Y(x)Q^{-1}\delta\beta|| + ||\int_a^b G(x, t)\delta(t)dt||. \qquad (10.12)$$

After some algebraic manipulation we obtain:

$$\max_{a \le x \le b} ||u(x) - y(x)|| \le \kappa_1 ||\delta\beta|| + \kappa_2 \max_{a \le x \le b} ||\delta(x)||, \qquad (10.13)$$

and

$$\max_{a \le x \le b} ||u(x) - y(x)|| \le \kappa \max(||\delta\beta||, \max_{a \le x \le b} ||\delta(x)||), \qquad (10.14)$$

where

$$\kappa_1 = \max_{a \le x \le b} ||Y(x)Q^{-1}||, \qquad \kappa_2 = sup_x \int_a^b ||G(x,t)|| dt, \qquad (10.15)$$

and

$$\kappa = \max_{a \le x \le b} (||Y(x)Q^{-1}|| + \int_a^b ||G(x,t)|| dt). \qquad (10.16)$$

Following the same procedure as above and using the 1-norm we obtain the corresponding parameters called $\gamma_1$, $\gamma_2$ and $\gamma$. For many problems of interest the relative sizes of the two parameters $\kappa_1$ and $\gamma_1$ tell us about the conditioning and the stiffness of the continuous problem.

Another important parameter is $\sigma$, which is called the "stiffness ratio". It is defined for linear problems (10.9) as

$$\sigma = \max \frac{\max_{a \le x \le b} ||u(x) - y(x)||}{\delta\beta \quad \dfrac{\int_a^b ||u(x) - y(x)|| dx}{(b-a)}}. \qquad (10.17)$$

where $u(x)$ is the solution of the perturbed equation with $\delta(x) \equiv 0$ (see [14,31]). If $\sigma$ is large we are dealing with problems possessing different time scales for which the growth or decay rates of some fundamental solution modes are very rapid compared to others. These parameters can be used to detect stiffness for initial and boundary value problems, see [7,8,23,26] for details.

## 10.2.5   Singular Problems

Consider the equation

$$y' = x^{-\alpha} F(x,y), \quad 0 < x < 1. \qquad (10.18)$$

Clearly we will have problems estimating the derivative $y'$ at the initial point of the integration interval $x = 0$. As a result this problem is singular. If $\alpha = 1$ it is said to have a singularity of the first kind; for $\alpha > 1$ it has a singularity of the second kind. A difficulty in solving these particular equations will arise only if the derivative needs to be evaluated at the boundary point $x = 0$.

Equations such as (10.18) arise for example when we reduce a PDE to an ODE using spherical or cylindrical symmetry (see Sect. 8.1.2). Typically these have as boundary condition that the gradient at $x = 0$ equals 0, so the problem of singularity is avoided. For a much more thorough discussion of singular problems the reader is referred to [5, p. 484].

## 10.3   Boundary Conditions

### 10.3.1   Separated Boundary Conditions

When defining the general form of a BVP (10.1), the boundary conditions are specified as a general function $g$ defined at the two boundary points $x = a$ and $x = b$. In this formalism it is possible to specify a condition which simultaneously involves both points.

There exists a simpler set of problems, where the boundary condition $g$ for any component is either given at $x = a$ or at $x = b$, and none of the boundary conditions is a function of both ends of the range of integration simultaneously.

Thus for each $1 \leq i \leq m$ either the $i$th row of $B_a$ or the $i$th row of $B_b$ are identically zero where, assuming that the equation is linear, we have

$$B_a = \frac{\partial g}{\partial y(a)} \quad \text{and} \quad B_b = \frac{\partial g}{\partial y(b)}, \tag{10.19}$$

so that the boundary value problem becomes:

$$y' = f(x,y), \quad a \leq x \leq b$$
$$B_a y(a) + B_b y(b) = rhs. \tag{10.20}$$

This is called the separated form of the boundary conditions.

Often we will convert a problem with non-separated boundary conditions to one with separated boundary conditions before finding an algorithm for its solution. A given BVP which has boundary conditions imposed in non-separated form can be converted to separated form by adding one trivial ODE per non-separated boundary condition. An account of how this can be carried out is given in [5, p. 6]. This technique can be extended to deal with the possibility that the boundary conditions are only partly separated or with multipoint boundary conditions.

We will give many examples of how a problem can be converted to standard form in the next chapter when we discuss R implementations.

## 10.3.2 Defining Good Boundary Conditions

To illustrate the fact that not all boundary conditions lead to a stable solution, we now consider a case where there is a dichotomy in the solution space. In particular we consider the following illuminating example of [35]:

$$y''' + 2y'' - y' - 2y = 0, \quad x \in [0, \infty]. \tag{10.21}$$

The general (analytic) solution to this equation is

$$y(x) = Ae^x + Be^{-x} + Ce^{-2x}. \tag{10.22}$$

This equation has one solution which increases as $x$ increases (related to the first term) and two which decay as $x$ increases. So, to obtain a dichotomy (see Sect. 10.2.3) we need to impose two boundary conditions at $x = 0$ and one at $x = \infty$. Suppose we impose

$$y(0) = 1, y'(0) = 1, y(\infty) = 0. \tag{10.23}$$

The boundary condition at $\infty$ gives: $A = 0$. The other two conditions imposed at $x = 0$ give $1 = B + C$ and $1 = -B - 2C$, so we obtain that $C = -2, B = 3$ and the solution with these particular boundary conditions is given by:

$$y(x) = 3e^{-x} - 2e^{-2x}, \tag{10.24}$$

and the solution will tend to 0 as $x \to \infty$. An interesting discussion of what happens when the wrong boundary conditions are imposed is given in [5, p. 117].

## 10.3.3 Problems Defined on an Infinite Interval

The above example had boundary conditions defined at $\infty$ and the problem was sufficiently simple to allow us to compute an analytic solution. In many cases we will need to resort to numerical techniques. In these cases, special care will be needed to deal with the boundary condition at $\infty$. One way of dealing with a problem defined on an infinite interval is to transform the independent variable, such that the new problem is defined on a finite region of integration.

Consider the simple problem

$$y' = F(x, y), \quad a \le x < \infty. \tag{10.25}$$

This can be mapped onto the region [0,1] by the transformation $t = a/x$, giving

$$t^2 \frac{dy}{dt} = -aF(a/t, y), \quad 0 \le t \le 1. \tag{10.26}$$

This is an ordinary differential equation with a singularity of the second kind.

## 10.4   Methods of Solution

There are two distinct numerical methods used to solve BVPs. Conceptually the simplest one which is the *shooting* method converts the BVP to an IVP. This is integrated with an appropriate IVP method while at the same time trying to solve for the values of the missing initial conditions. This algorithm is very easy to use but it has the disadvantage that it can suffer severe stability problems.

Alternatively, the equation can be approximated by *finite differences* in a way similar to the approximation of partial differential equations considered in Chap. 8.

## 10.5   Shooting Methods for Two-Point BVPs

### 10.5.1   The Linear Case

It is instructive to outline the procedure of shooting first for a *linear* second order equation.

$$y'' + c(x)y' + d(x)y = g(x), \quad a \le x \le b, \tag{10.27}$$

subject to the boundary conditions:

$$y(a) = \alpha \tag{10.28}$$

$$y(b) = \beta. \tag{10.29}$$

Finding a solution using shooting proceeds in several steps. First of all instead of using the given boundary conditions we replace the last one (10.29) with the *initial* condition

$$y'(a) = \gamma. \tag{10.30}$$

Of course, the value of $\gamma$ is unknown and we need to calculate it to find the solution.

We now solve the initial value problem (10.27), subject to the initial conditions (10.28) and (10.30) using one of the IVP solvers discussed in previous chapters. We denote the solution of this initial value problem by $w(x)$.

This solution produces an estimate of $y$ at the end of the integration interval, $b$, which we can compare with the imposed boundary condition (10.29). We now compute a second approximate solution with the initial condition

$$y'(a) = \delta. \tag{10.31}$$

We denote this solution by $r(x)$. We now take a linear combination of both solutions, of the form:

$$y(x) = \theta w(x) + (1 - \theta)r(x), \tag{10.32}$$

so that $y(x)$ is the required solution. Assuming that $w(b) \neq r(b)$, we choose the value of $\theta$ as:

$$\theta = \frac{\beta - r(b)}{w(b) - r(b)}. \tag{10.33}$$

As in this case

$$\beta = \theta w(b) + (1 - \theta)r(b), \tag{10.34}$$

it is easy to see that (10.32) and (10.33) satisfy (10.27).

## 10.5.2  The Nonlinear Case

The shooting algorithm is not so easy to use if the problem is nonlinear. To show this we consider the general nonlinear second order equation

$$y'' + f(x, y, y') = 0, \quad a \leq x \leq b, \tag{10.35}$$

subject to the boundary conditions

$$y(a) = \alpha \tag{10.36}$$

$$y(b) = \beta. \tag{10.37}$$

We assume that these equations have a unique solution $y(x)$.

Following the procedure considered in the linear case, we replace the last boundary value, (10.37), with the *initial* condition

$$y'(a) = s. \tag{10.38}$$

The value of $s$ is unknown, and we need to estimate it.

Starting with a first guess for $s$, we now solve the problem (10.35), subject to the initial conditions (10.36) and (10.38). We denote the solution of this initial value problem by $y(x; s)$. This solution produces an estimate of $y$ at the end of the integration interval and we can compare this with the imposed boundary condition (10.37). Our aim now is to find a value of $s$, say $s^*$ so that the second boundary condition (10.37) is satisfied, i.e. $y(b; s^*) = \beta$.

In contrast to the linear case, finding the value of $s^*$ requires the solution of a nonlinear algebraic equation and it is this that makes the nonlinear problem considerably harder to solve than the linear one. This nonlinear solution algorithm is very well known and is described for example in [5, p. 134]. Note that this approach can also be extended to deal with more general boundary conditions of the form $g(y(a), y(b)) = 0$.

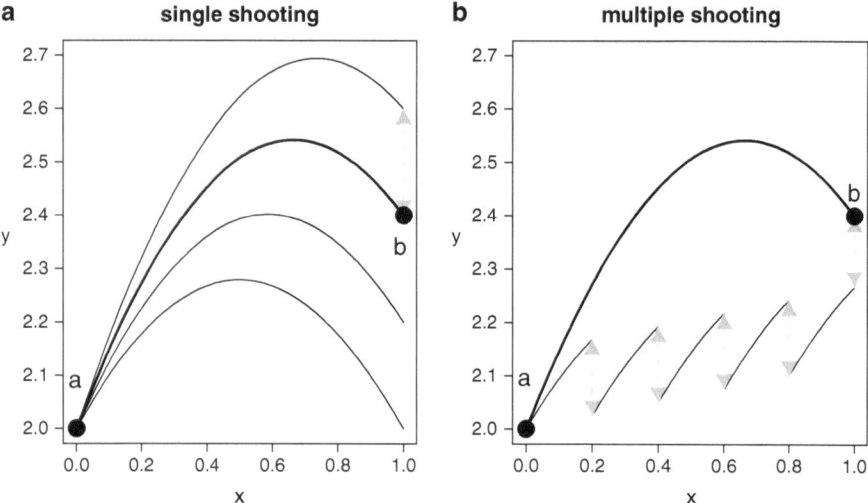

**Fig. 10.1** Shooting methods applied to solve a second order boundary value problem on the interval [0,1] with boundary conditions $y(0) = a, y(1) = b$. (**a**) In the single shooting method the equation is solved with an IVP method based on a guess of the unknown initial condition $y'(0)$. A Newton-Raphson method is applied to search for the values of $y'(0)$ such that $y(1) = b$. (**b**) In the multiple shooting method, the integration interval is divided in small subintervals, and a Newton-Raphson method is used to solve for the unknown initial values in each of the subintervals

The algorithm described above is called *simple shooting* (see Fig. 10.1a). As simple shooting often suffers from a severe loss of stability, it needs to be used with care. However, it is such a natural and easy to understand algorithm, that many users try this algorithm first of all. If it fails, one needs to revert to a better algorithm such as multiple shooting or finite differences as described in the next sections.

## 10.5.3 Multiple Shooting

The big problem with simple shooting is that it can be set up so that we are solving unstable IVPs. This gives rise to very severe error accumulation. This is made worse if we increase the range of integration, so a natural solution is to divide the mesh into a number of smaller components, an approach called *multiple shooting* (see Fig. 10.1b).

For example, we could subdivide the interval $[a,b]$ into $a = x_0 < x_1 < \ldots < x_N = b$. An initial value problem is then integrated over each interval and these solutions are joined together to form a continuous solution. It is the fact that these integrations are performed over relatively short intervals that is important, since it does not give the error "time" to accumulate. The concept of multiple shooting is very well known and is described for example in [5, p. 145], while a multiple shooting algorithm is given in [5, p. 517].

## 10.6   Finite Difference Methods

Finite difference methods approximate the solution of the differential equation on the entire interval by approximating the derivative using difference schemes. For this reason they are called "global methods" in the literature. Consider again problem (10.20). The first step is to divide the range $[a,b]$ into a mesh

$$\pi : \{a = x_0, x_1, \ldots, x_N = b\}, \qquad (10.39)$$

and we define $h_i = x_i - x_{i-1}$, $1 \leq i \leq N$. Then we approximate the derivatives at each point in the mesh using a finite difference scheme. This produces a set of algebraic equations, the solution of which gives an approximation to the solution of (10.20) at the mesh points defined in $\pi$. It is important to realize that not all difference schemes are useful because, for boundary value problems, there is no preferred direction of integration, which is in contrast to initial value problems where the direction is imposed by the "initial" value. In fact, an important property that a numerical scheme for the solution of BVPs should satisfy is that of so-called "Time Reversal Symmetry", see [9, Chap. 9]. This means that the scheme must provide the same discrete approximation on the interval $[a,b]$ when the variable $x$ of the continuous problem is transformed into $\xi = a+b-x$ and the boundary conditions are changed accordingly. This property is important because BVPs have both increasing and decreasing modes in the solution and the numerical method should integrate forward and backward without a preferential direction. The numerical schemes that have the time reversal symmetry property are called symmetric schemes (see [5, p. 440]) and it is natural to use symmetric methods, as such methods are "direction independent".

### 10.6.1   A Low Order Method for Second Order Equations

We consider first of all a particularly simple finite difference scheme applied to the two point boundary value problem

$$\frac{d^2 y}{dx^2} = f(x,y), \quad y(0) = \alpha,\ y(1) = \beta, \quad 0 < x < 1. \qquad (10.40)$$

As previously explained, we first choose a grid $\pi$ defined by (10.39) and, for the sake of clarity, we will assume that $h$ is constant and $a = 0$, $b = 1$. We wish to approximate the second derivative of $y$ at the mesh points $x_i = x_0 + ih$, $1 \leq i \leq N - 1$. To do this we make the approximations:

$$h^2 y''(x_i) \approx y_{i+1} - 2y_i + y_{i-1}, \qquad i = 1, \ldots, N-1. \qquad (10.41)$$

Substituting this into (10.40) we obtain

$$y_{i+1} - 2y_i + y_{i-1} = h^2 f(x_i, y_i), \qquad i = 1, \ldots, N-1, \qquad y_0 = \alpha,\ y_N = \beta. \qquad (10.42)$$

Equations (10.42) generate a tri-diagonal system of algebraic equations which defines $y_i$, $1 \leq i \leq N-1$ and this can be solved to give a first discrete approximation to the solution of (10.40). This technique is a very cheap but not an especially powerful one. It can be useful if only a low order of accuracy is required or if the solution is very smooth but perhaps one of the main reasons for considering it is that it extends directly to the solution of some important classes of partial differential equations (e.g. Sect. 8.3.1). A variable stepsize implementation of this technique is described in [28, 29].

## 10.6.2   Other Low Order Methods

In this section we again consider the numerical solution of the system of $m$ Two-Point Boundary Value Problems.

$$y' = f(x,y), \quad a \leq x \leq b$$
$$g(y(a), y(b)) = 0, \quad y \in \mathfrak{R}^m. \tag{10.43}$$

The two most important one step symmetric schemes that can be used to solve BVPs are the implicit midpoint rule and the trapezoidal rule. Let's consider the discrete approximation given by the implicit midpoint rule. If we apply this to (10.43) we obtain

$$\frac{y_n - y_{n-1}}{h_n} = f(x_{n-1/2}, \frac{1}{2}(y_n + y_{n-1})), \quad n = 1, 2, \ldots, N$$
$$g(y_0, y_N) = 0. \tag{10.44}$$

These equations now define $m(N+1)$ algebraic equations in $m(N+1)$ unknowns. Having solved these equations we obtain a discrete approximation to $y(x)$ on the grid $\pi$. There exist many ways to solve these equations, the most well-known one being a variant of Newton's method [5]. To show how this algorithm works, we consider the rather more simple:

$$y' = A(x,y)y + c(x), \quad a \leq x \leq b$$
$$g(y(a), y(b)) = 0. \tag{10.45}$$

To make the problem even more simple we assume that $A$ is independent of $y$, and we use simplified boundary conditions

$$y' = A(x)y + c(x), \quad a \leq x \leq b$$
$$B_a y(a) + B_b y(b) = rhs. \tag{10.46}$$

Applying the implicit midpoint rule to this equation, we obtain

$$\frac{y_n - y_{n-1}}{h_n} = A(x_{n-1/2}) \left( \frac{y_n + y_{n-1}}{2} \right) + c(x_{n-1/2}), \quad n = 1, 2, \ldots, N \tag{10.47}$$
$$B_a y_0 + B_b y_N = rhs.$$

This system of $m(N+1)$ algebraic equations can be written as:

$$M y_\pi = \Gamma \tag{10.48}$$

$$M = \begin{bmatrix} S_1 & R_1 & & & \\ & S_2 & R_2 & & \\ & & \ddots & \ddots & \\ & & & S_N & R_N \\ B_a & & & & B_b \end{bmatrix}, \quad y_\pi = \begin{bmatrix} y_0 \\ y_1 \\ \vdots \\ y_N \end{bmatrix}, \quad \Gamma = \begin{bmatrix} c(x_{1/2}) \\ c(x_{3/2}) \\ \vdots \\ c(x_{N-1/2}) \\ rhs \end{bmatrix}, \tag{10.49}$$

where $S_n = -[h_n^{-1} I + A(x_{n-1/2})/2]$, $R_n = [h_n^{-1} I - A(x_{n-1/2})/2]$.

The important property that this system of algebraic equations possesses is that it is almost tridiagonal so it is relatively cheap to solve. An algorithm to solve this equation is discussed e.g. in [5, Chap. 7]. The extension to nonlinear equations is conceptually easy, but often much more difficult to implement because the convergence of Newton's method is not guaranteed (see [5, Chap. 5]).

If we apply the trapezoidal rule to (10.46) we obtain:

$$\frac{y_n - y_{n-1}}{h_n} = \frac{1}{2} (A(x_n) y_n + A(x_{n-1}) y_{n-1} + c(x_n) + c(x_{n-1})), \quad n = 1, 2, \ldots, N$$
$$B_a y_0 + B_b y_N = rhs. \tag{10.50}$$

The resulting linear system has a matrix with the same sparsity structure as system (10.49) but now $S_n = -h_n^{-1} I - \frac{1}{2} A(x_{n-1})$, $R_n = h_n^{-1} I - \frac{1}{2} A(x_n)$ and $\Gamma_n = \frac{1}{2}(c(x_{n-1}) + c(x_n))$

The midpoint rule and the trapezoidal rule, which are both of second order give second order convergence when applied to a boundary value problem. As mentioned previously the low order methods are only useful when a low degree of accuracy is required or if the solution is very smooth. When higher accuracy is required, it is normally more efficient to use higher order numerical algorithms.

### 10.6.3 Higher Order Methods Based on Collocation Runge-Kutta Schemes

An obvious way of deriving high order numerical methods is simply to approximate the derivative using high order Runge-Kutta formulae. The requirement that the method should be symmetric leads us to consider Gauss or Lobatto points in our numerical algorithm. The most widely used implicit Runge-Kutta methods for the solution of BVPs are "collocation methods". The idea behind collocation is to find a set of polynomials $\phi_n(x)$ of degree $s$ that satisfies the differential problem in the interval $[x_{n-1}, x_n]$ on a set of collocation points $x_{ni} = x_{n-1} + c_i h_n, i = 1, \ldots, s, \quad n = 1, \ldots, N$, with $0 < c_0 < c_1 < \cdots < c_s < 1$ distinct real numbers. The polynomial $\phi_n(x)$ satisfies

$$\begin{aligned} \phi_n(x_{n-1}) &= y_{n-1} \\ \phi'_n(x_{ni}) &= f(x_{ni}, \phi(x_{ni})), \quad i = 1, 2, \ldots, s, \end{aligned} \tag{10.51}$$

and the numerical approximation at $x_n$ is then given by $y_n = \phi_n(x_{n-1} + h_n)$. If the $c_i, i = 1, \ldots, s$ are distinct then the collocation method in each interval is equivalent to an $s$ stage implicit Runge-Kutta method, where the $c_i$ are the stages of the Runge-Kutta scheme (see [21, Theorems 7.7,7.8], [5, p. 210]). The implicit midpoint rule (10.44) is a second order collocation Runge-Kutta method with $c_1 = 1/2$ and collocation points $x_{n1} = x_{n-1} + c_1 h_n, n = 1, \ldots, N$.

A popular way of specifying the collocation points is to use Gauss points (see [5, 11]) since these are in some sense optimal in that the maximum order of an $s$-stage Gauss method is $2s$. This is the choice used in the well known collocation code COLSYS (see [5]). In [11] there is a detailed description of how a fourth order Gauss Runge-Kutta method can be implemented to solve a general two-point boundary value problem.

## 10.6.4   Higher Order Methods Based on Mono Implicit Runge-Kutta Formulae

An alternative class of Runge-Kutta methods which have a smaller linear algebra cost than collocation methods are mono-implicit Runge-Kutta (MIRK) methods. These were originally developed as backward versions of Runge-Kutta methods and they can be obtained by replacing $h$ by $-h$ in an explicit Runge-Kutta formula (2.3).

For example, if we consider the explicit Runge-Kutta method

$$y_{n+1} - y_n = h/2[f(y_n) + f(y_n + h f(y_n))], \tag{10.52}$$

and replace $h$ by $-h$ we obtain

$$\begin{aligned} y_{n-1} - y_n &= -h/2[f(y_n) + f(y_n - h f(y_n))] \\ &\text{or} \\ y_n - y_{n-1} &= h/2[f(y_n) + f(y_n - h f(y_n))]. \end{aligned} \tag{10.53}$$

The important property of (10.52) and (10.53) is that the stability region of the former is the reflection of the stability region of the latter in the imaginary axis. In particular MIRK methods have been described in some detail in [10, 15, 19], and these methods are implicit in the single unknown $y_{n+1}$. For example the fourth order Lobatto IIIa formula

$$
\begin{array}{c|ccc}
0 & 0 & 0 & 0 \\
1/2 & 5/24 & 1/3 & -1/24 \\
1 & 1/6 & 2/3 & 1/6 \\
\hline
 & 1/6 & 2/3 & 1/6
\end{array}
\tag{10.54}
$$

can be written as:

$$
y_{n+1} - y_n = \tfrac{h}{6}(k_1 + 4k_2 + k_3)
$$
where
$$
\begin{aligned}
k_1 &= f(x_n, y_n) \\
k_2 &= f(x_{n+1/2}, \tfrac{y_n + y_{n+1}}{2} - \tfrac{h}{8}(k_3 - k_1)) \\
k_3 &= f(x_{n+1}, y_{n+1}).
\end{aligned}
\tag{10.55}
$$

It can immediately be seen that the only unknown in this re-formulation of the Lobatto method is $y_{n+1}$.

Once we have obtained an approximate solution defined on a discrete grid $\pi$ it is often the case that we wish to derive a continuous solution so that we can obtain an approximation to the solution at non-grid points. A technique for deriving continuous MIRK formulae has been derived in [33] where these continuous formulae are called CMIRK formulae.

### 10.6.5   Higher Order Methods Based on Linear Multistep Formulae

Another way to derive high order methods is to use linear multistep formulae of high order in a boundary value approach deriving a set of finite difference formulas. In this case the numerical scheme on the grid of $N+1$ mesh points $x_0 < x_1 < \cdots < x_N$ generates the following discrete problem:

$$
\left\{
\begin{aligned}
& g(a,b) = 0 \\
& \sum_{i=0}^{k} \alpha_i^{(n)} y_i = h_n \sum_{i=0}^{k} \beta_i^{(n)} f_i, && n = 1, \ldots, k_1 - 1, \\
& && \text{(additional initial methods)} \\
& \sum_{i=0}^{k} \alpha_i^{(n)} y_{n-k_1+i} = h_n \sum_{i=0}^{k} \beta_i^{(n)} f_{n-k_1+i}, \; n = k_1, \ldots, N - k_2, \\
& && \text{(main method)} \\
& \sum_{i=0}^{k} \alpha_i^{(n)} y_{N-k+i} = h_n \sum_{i=0}^{k} \beta_i^{(n)} f_{N-k+i}, \; n = N+1 - k_2, \ldots, N, \\
& && \text{(additional final methods).}
\end{aligned}
\right.
\tag{10.56}
$$

The coefficients $\alpha_i^{(n)}, \beta_i^{(n)}, \ i = 0, k, \ \ n = 1, \ldots, N$ are computed using a variable coefficient technique (see Sect. 2.2.4.2). These methods, called Boundary Value Methods (see Sect. 2.3), have been successfully used for the solution of BVPs, in particular the two classes called Top Order Methods and BS schemes, both of which are symmetric and widely used. The latter schemes define on $[a, b]$ a spline function $S(x)$ that collocates the numerical solution at the mesh-points $x_i$, that is $S'(x_i) = f(x_i, S(x_i)), \ \ i = 0, 1, \ldots, N$. The collocation solution is continuous up to the $k$th derivative. The trapezoidal rule is the simplest method of both classes.

## 10.6.6   Deferred Correction

In this section we briefly consider the approach of iterated deferred correction for the numerical solution of BVPs. Deferred correction is a very widely used procedure originally proposed by Fox [20].

The idea behind it is to use a simple low order method to compute an initial solution of the equation and then solve a series of correction equations each of which improves the accuracy of the provisional solution. Thus we construct two differential operators one, $\Phi$, which is a cheap low order method used to compute a first approximation to the solution and another operator $\Psi$ which estimates the local error in $\Phi$. A simple deferred correction scheme has the form:

$$\begin{aligned} \Phi(\eta) &= 0 \\ \Phi(\overline{\eta}) &= \Psi(\eta). \end{aligned} \tag{10.57}$$

There are several ways in which the operators $\phi$ and $\psi$ can be defined. As already explained, we wish to choose $\phi$ so that it is easy to solve the problem $\phi(\eta) = 0$ for $\eta$. In choosing $\psi$ we want a method for which it is easy to compute $\psi(\eta)$ cheaply. Since MIRK formulae are implicit in a single unknown it turns out that the choice where $\phi$ and $\psi$ are fourth order and sixth order MIRK formulae has several important computational advantages. There are a couple of ways that could be used to obtain still higher order formulae. One approach is that adopted by [24] where high order formulae are obtained by approximating terms in the expansion of the local truncation error of the trapezoidal rule. However problems arise near both ends of the mesh and this has become known as the end of the net catastrophe. Another approach is to use rather more stable formulae where $\phi$ and $\psi$ are fourth order and sixth order Lobatto Runge-Kutta formulae respectively. Finally we can obtain higher order formulae by extending (10.57) to a formula of the form

$$\begin{aligned} \Phi(\eta) &= 0 \\ \Phi(\overline{\eta}) &= \Psi(\eta) \\ \Phi(\overline{\overline{\eta}}) &= \alpha(\eta) + \beta(\overline{\eta}). \end{aligned} \tag{10.58}$$

By choosing the operators appearing in (10.58) in a careful way we can derive a deferred correction method of order 8. Guidance on how these operators are chosen can be found in [25, 37].

## 10.7 Codes for the Numerical Solution of Boundary Value Problems

Two of the earliest and most widely used codes for the solution of boundary value problems are COLSYS [4] and COLNEW [6]. These are codes that implement Runge-Kutta collocation methods, and differ only in the type of spline used for representing the numerical solution (B-spline versus monomial splines), and the linear algebra solvers.

Two widely used codes which are based on deferred correction schemes are TWPBVP and TWPBVPL [13, 16]. They use MIRK schemes and Lobatto formulae respectively [22, p. 75]. Both codes have been extended with a mesh selection based on conditioning (TWPBVPC, TWPBVPLC) [12]. The Lobatto formulae have much better stability than the MIRK formulae, but have more computational overhead. As a result TWPBVP is efficient for non stiff or mildly stiff problems, whereas TWPBVPL is efficient for very stiff boundary value problems.

The codes COLMOD [17] and ACDC [17], which are based on collocation and MIRK formulae respectively, implement an automatic continuation strategy and these are often much more efficient on extremely stiff problems than codes which do not allow continuation.

All codes can solve problems with non-separated boundary conditions, but only COLSYS and COLNEW accept multipoint BVPs, as well as higher order systems.

MIRK methods are also the underlying formulae of the codes MIRKDC [19], BVP_SOLVER [36] and bvp4c [34].

Symmetric BVMs have been implemented in the code TOM with a mesh selection based on conditioning and a quasi-linearization strategy for the solution of the non linear equations [27, 30, 32].

For the numerical solution of boundary value differential-algebraic equations the code COLDAE [2] is available.

Finally, MUSN [5] is a code based on multiple shooting.

The codes COLNEW [6], COLSYS [3], COLMOD [17], TWPBVP [16], TWPBVPC [12], TWPBVPLC [13] and ACDC [17] are implemented in R (see Sect. A.3).

## References

1. Ascher, U. M., & Petzold, L. R. (1998). *Computer methods for ordinary differential equations and differential-algebraic equations*. Philadelphia: SIAM.
2. Ascher, U. M., & Spiteri, R. J. (1994). Collocation software for boundary value differential-algebraic equations. *SIAM Journal on Scientific Computing, 15*(4), 938–952.

3. Ascher, U. M., Christiansen, J., & Russell, R. D. (1979). **COLSYS**–a collocation code for boundary value problems. In B. Childs et al. (Ed.), *Lecture notes in computer science 76* (pp. 164–185). New York: Springer.
4. Ascher, U. M., Christiansen, J., & Russell, R. D. (1981). Collocation software for boundary-value ODEs. *ACM Transactions on Mathematical Software, 7*, 209–222.
5. Ascher, U. M., Mattheij, R. M. M., & Russell, R. D. (1995). *Numerical solution of boundary value problems for ordinary differential equations.* Philadelphia: SIAM.
6. Bader, G., & Ascher, U. M. (1987). A new basis implementation for a mixed order boundary value ODE solver. *SIAM Journal on Scientific and Statistical Computing, 8*, 483–500.
7. Brugnano, L., Mazzia, F., & Trigiante, D. (2011). Fifty years of stiffness. In T. E. Simos (Ed.), *Recent advances in computational and applied mathematics.* Dordrecht/New York: Springer.
8. Brugnano, L., & Trigiante, D. (1996). On the characterization of stiffness for ODEs. *Dynamics of Continuous, Discrete and Impulsive Systems, 2*(3), 317–335.
9. Brugnano, L., & Trigiante, D. (1998). *Solving differential problems by multistep initial and boundary value methods: Vol. 6. Stability and control: Theory, methods and applications.* Amsterdam: Gordon and Breach.
10. Cash, J. R. (1975). A class of implicit Runge–Kutta methods for the numerical integration of stiff ordinary differential equations. *Journal of Alternative and Complementary Medicine, 22*, 504.
11. Cash, J. R. (2004). A survey of some global methods for solving two-point boundary value problems. *Applied Numerical Analysis and Computational Mathematics, 1*, 1–17.
12. Cash, J. R., & Mazzia, F. (2005). A new mesh selection algorithm, based on conditioning, for two-point boundary value codes. *Journal of Computational and Applied Mathematics, 184*, 362–381.
13. Cash, J. R., & Mazzia, F. (2006). Hybrid mesh selection algorithms based on conditioning for two-point boundary value problems. *Journal of Numerical Analysis, Industrial and Applied Mathematics, 1*(1), 81–90.
14. Cash, J. R., & Mazzia, F. (2009). Conditioning and hybrid mesh selection algorithms for two-point boundary value problems. *Scalable Computing: Practice and Experience, 10*(4), 347–361.
15. Cash, J. R., & Singhal, A. (1982). High order methods for the numerical solution of two-point boundary value problems. *BIT, 22*, 184.
16. Cash, J. R., & Wright, M. H. (1991). A deferred correction method for nonlinear two-point boundary value problems: Implementation and numerical evaluation. *SIAM journal on scientific and statistical computing, 12*, 971–989.
17. Cash, J. R., Moore, G., & Wright, R. W. (1995). An automatic continuation strategy for the solution of singularly perturbed linear two-point boundary value problems. *Journal of Computational Physics, 122*, 266–279.
18. Davis, H. T. (1962). *Introduction to nonlinear differential and integral equations.* New York: Dover.
19. Enright, W. H., & Muir, P. (1980). Efficient classes of Runge–Kutta methods for two point boundary value problems. *Computing, 37*, 315.
20. Fox, L. (1957). *The numerical solution of two point boundary value problems in ordinary differential equations.* London: Clarendon Press.
21. Hairer, E., Norsett, S. P., & Wanner, G. (2009). *Solving ordinary differential equations I: Nonstiff problems. Second revised edition.* Heidelberg: Springer.
22. Hairer, E., & Wanner, G. (1996). *Solving ordinary differential equations II: Stiff and differential-algebraic problems.* Heidelberg: Springer.
23. Iavernaro, F., Mazzia, F., & Trigiante, D. (2006). Stability and conditioning in numerical analysis. *Journal of Numerical Analysis, Industrial and Applied Mathematics, 1*(1), 91–112.
24. Lentini, M., & Pereyra, V. (1977). An adaptive finite difference solver for nonlinear two point boundary value problems with mild boundary layers. *SIAM Journal of Numerical Mathematics, 14*, 91–111.

25. Lindberg, B. (1980). Error estimation and iterative improvement for discetization algorithms. *BIT, 20*, 486.
26. Mazzia, F., & Nagy, A. M. (2010). Stiffness detection strategy for explicit Runge–Kutta methods. *AIP Conference Proceedings, 1281*(1), 239–242.
27. Mazzia, F., & Sgura, I. (2002). Numerical approximation of nonlinear BVPs by means of BVMs. *Applied Numerical Mathematics, 42*(1–3), 337–352. Ninth Seminar on Numerical Solution of Differential and Differential-Algebraic Equations (Halle, 2000).
28. Mazzia, F., & Trigiante, D. (1992). Numerical methods for second order singular perturbation problems. *Computers and Mathematics with Applications, 23*(11), 81–89.
29. Mazzia, F., & Trigiante, D. (1993). Numerical solution of singular perturbation problems. *Calcolo, 30*(4), 355–369 (1995).
30. Mazzia, F., & Trigiante, D. (2004). A hybrid mesh selection strategy based on conditioning for boundary value ODE problems. *Numerical Algorithms, 36*(2), 169–187.
31. Mazzia, F., & Trigiante, D. (2010). Efficient strategies for solving nonlinear problems in BVPs codes. *Nonlinear Studies, 17*(4), 309–326.
32. Mazzia, F., Sestini, A., & Trigiante, D. (2009). The continuous extension of the B-spline linear multistep methods for BVPs on non-uniform meshes. *Applied Numerical Mathematics, 59*(3–4), 723–738.
33. Muir, P. H., & Owren, B. (1993). Order barriers and charaterisations of continuous mono-implicit Runge–Kutta schemes. *Mathematics of Computation, 61*, 675.
34. Shampine, L. F., Kierzenka, J., & Reichelt, M. W. (2000). *Solving boundary value problems for ordinary differential equations in MATLAB with bvp4c.* In Matlab Guide, D.J. Higham and N.J. Higham, pp 163–169, Philadelphia: SIAM.
35. Shampine, L. F., Gladwell, I., & Thompson, S. (2003). *Solving ODEs with MATLAB.* Cambridge: Cambridge University Press.
36. Shampine, L. F., Muir, P. H., & Xu, H. (2006). A user-friendly Fortran BVP solver. *Journal of Numerical Analysis, Industrial and Applied Mathematics, 1*(2), 201–217.
37. Skeel, R. D. (1982). A theoretical framework for providing accurracy results for deferred corrections. *SINUM Journal on Numerical Analysis, 19*, 171–196.
38. Strang, G., & Fix, G. (1973). *Analysis of the finite element method.* Englewood Cliffs, NJ: Prentice Hall.

# Chapter 11
# Solving Boundary Value Problems in R

**Abstract** Boundary Value Problems can be solved in R using shooting, MIRK and collocation methods and these can be found in the R package **bvpSolve** . The functions in this R package have an interface which is similar to the interface of the initial value problem solvers in the package **deSolve**. The default input to the solvers is very simple, requiring specification of only one function that calculates the derivatives while the boundary conditions are represented as simple vectors. However, in order to speed-up the simulations, and to increase the number of problems that can be solved, it is also possible to specify the boundary conditions by means of a function and provide analytic functions for the derivative and boundary gradients. In this chapter we demonstrate how to solve BVPs using a variety of well-known test problems, illustrating a wide range of difficulties in solving BVPs. We show how to use (manual and automatic) continuation, how difficult boundary conditions can be handled, and give many examples of how to convert BVPs to standard form. Some BVPs are much better solved using the finite difference methods as explained in the PDE chapter. We give an example of such a boundary value problem at the end of this chapter.

## 11.1  Boundary Value Problem Solvers in R

There are three functions to solve boundary value problems in the R package **bvpSolve** [20]. They implement a shooting method (bvpshoot), using the solvers from the R package **deSolve**, a MIRK method (bvptwp), based on the code TWPBVP [11], TWPBVPC [8], TWPBVPL [9] and ACDC [12] and a collocation method (bvpcol), implementing the codes COLNEW [6], COLSYS [3] and COLMOD [11]. Their (simplified) syntax is:

```
bvpshoot(yini, x, func, yend, parms, order, ...)
bvptwp(yini, x, func, yend, parms, order, ...)
bvpcol(yini, x, func, yend, parms, order, ...)
```

K. Soetaert et al., *Solving Differential Equations in R*, Use R!,
DOI 10.1007/978-3-642-28070-2_11, © Springer-Verlag Berlin Heidelberg 2012

where `func` is the derivative function and `parms` is the parameter `vector` or `list`. These two arguments are similar to the case of initial value problem solvers from the R package **deSolve**. However, the independent variable is called `x` here, rather than `times` for IVPs. The boundary conditions are specified by `yini` and `yend`, for the first and last point respectively. They are both a vector with length equal to the number of dependent variables, and having `NA` where the boundary value is not known. A more flexible (but more cumbersome) way of defining boundary conditions, by a function, is also available. The default tolerances are $10^{-8}$ for all the codes. The codes `bvptwp` and `bvpcol` require only one input error tolerance.

Functions `bvpshoot` and `bvptwp` can solve two-point boundary value problems only while `bvpcol` also finds solutions for multipoint problems.

As `bvptwp` and `bvpcol` generally lead to more accurate solutions, and provide solutions where `bvpshoot` fails, the latter should normally not be used, unless to find good initial guesses for `bvpcol` or `bvptwp`.

We can choose as polynomial basis either B-splines or monomial splines (`bvpcol`), MIRK or Lobatto (`bvptwp`). Both functions also include an automatic continuation strategy. See appendix, Table A.9.

In addition to the solvers, the **bvpSolve** package also contains specially-designed methods for visualization (`plot`) or to print the `diagnostics`. The syntax of the solvers and these functions is extensively illustrated in the following examples.

## 11.2   A Simple BVP Example

We illustrate the basic syntax of the BVP solvers by means of a simple BVP ODE (which is Problem 7 from the test problems available from [7]):

$$\varepsilon y'' + xy' - y = -(1 + \varepsilon \pi^2)\cos(\pi x) - \pi x \sin(\pi x)$$
$$y(-1) = -1 \qquad\qquad (11.1)$$
$$y(1) = 1.$$

As this is a second order ODE, the system is fully determined by the two boundary conditions. The parameter $\varepsilon$ will be used to illustrate the effect of stiffness of BVPs on their ease of solution. The smaller $\varepsilon$ is, the more stiff is the problem.

### 11.2.1   Implementing the BVP in First Order Form

If we expand the second order ODE as two first order ODEs, we obtain:

$$y' = y_1$$
$$y_1' = \frac{1}{\varepsilon}(-xy_1 + y - (1 + \varepsilon \pi^2)\cos(\pi x) - \pi x \sin(\pi x)), \qquad (11.2)$$

where $y_1$ is the first order derivative. This problem is implemented as:

```
prob7 <- function(x, y, pars) {
  list(c( y[2],
          1/eps * (-x*y[2] + y[1] - (1+eps*pi*pi)*
                   cos(pi*x) - pi*x*sin(pi*x))))
  }
```

where y is a 2-valued vector that contains $y$ (=y[1]) and $y_1$ (=y[2]). The problem is solved as:

```
library(bvpSolve)
eps  <- 0.1
sol <- bvptwp(yini = c(y = -1, y1 = NA),
              yend = c(1, NA), func = prob7,
              x = seq(-1, 1, by = 0.01))
```

Note how the boundary conditions at the start (yini) and at the end (yend) of the integration interval are specified, where NA ("not available") is used for unspecified boundary conditions. For the boundary conditions, we must provide the values of $y$ as well as of $y_1$, as this is a second order equation. Also, the dependent variables are given *names* in the specification of yini. This simplifies plotting the output (see below). Similar to what happens for initial value problems implemented in R , the derivative function must return the derivatives in the same order as that in which the initial and end conditions have been defined.

## 11.2.2  *Implementing the BVP in Second Order Form*

The function can also be implemented in second order form (11.1) as:

```
prob7_2 <- function(x, y, pars) {
  list(1/eps * (-x*y[2] + y[1] - (1+eps*pi*pi)*
                cos(pi*x) - pi*x*sin(pi*x)))
  }
sol1 <- bvptwp(yini = c(y = -1, y1 = NA),
               yend = c(1, NA), func = prob7_2,
               order = 2, x = seq(-1, 1, by = 0.01))
```

where the argument order = 2 sets the order of the differential equation. When implemented in second order form, we have to provide boundary values not just for $y$, the first element in yini or yend, but also for the first derivative of $y$, which is the second element in yini. The argument y passed to the function prob7_2 will contain the current values of $y$ (y[1]) and $y'$ (y[2]). We only return the second order derivative from the function prob7_2.

  With this value of the parameter $\varepsilon$ (eps), the solution curves are rather smooth (Fig. 11.1 solid line)

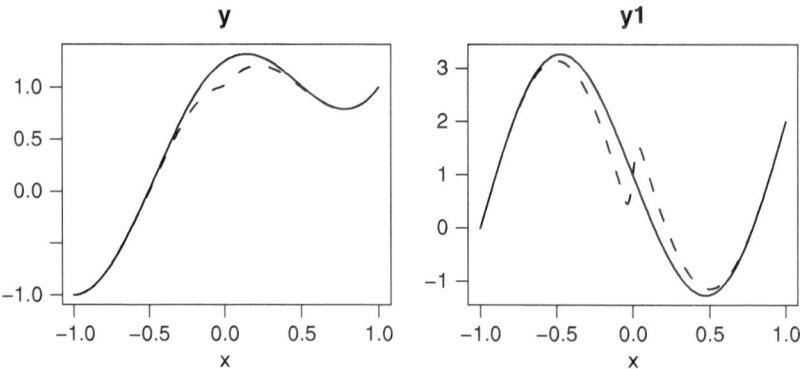

**Fig. 11.1** Solution of the test Problem 7, for $\varepsilon = 0.01$ (*solid line*) and $\varepsilon = 0.0001$ (*dotted line*). See text for the R code

As for IVP solvers, the solution matrix produced by the BVP solvers consists of several columns, the first of which holds the independent variable, x.

```
head(sol, n=3)
```

```
         x         y        y1
[1,]  -1.00  -1.0000000  0.001699844
[2,]  -0.99  -0.9994887  0.100558398
[3,]  -0.98  -0.9979891  0.199338166
```

For increasingly small values of the parameter $\varepsilon$ the problem becomes more and more difficult to solve due to the presence of a zone of very rapid change near $x = 0$. In fact, the shooting method completely fails for $\varepsilon = 0.0005$, while the other methods still give good solutions.

```
eps <-0.0005
sol2 <- bvptwp(yini = c(y = -1, y1 = NA),
               yend = c(1, NA), func = prob7,
               x =  seq(-1, 1, by=0.01))
```

We plot the results with **bvpSolve**'s plot method, which depicts both dependent variables. We plot the results of both problems together. Note the zone near 0 when $\varepsilon$ is very small (Fig. 11.1, dashed line)

```
plot(sol, sol2, col = "black", lty = c("solid", "dashed"),
     lwd = 2)
```

## 11.3   A More Complex BVP Example

It is not always necessary to manually translate higher order boundary value problems into multi-dimensional first order problems. As already illustrated in the first example we can avoid such manual manipulations by explicitly giving the order of the problem to the solver. However, in such cases we have to take care to provide all necessary boundary values and provide them in the correct order. We illustrate this with the test problem referred to as "swirling flow III" [5], which was briefly discussed in Sect. 1.1.3. We also use this problem to illustrate that there are limits on the precision that can be obtained.

The problem

$$g'' = (gf' - fg')/\varepsilon$$
$$f'''' = (-ff''' - gg')/\varepsilon, \qquad (11.3)$$

is defined on the interval [0,1] and subject to boundary conditions:

$$g(0) = -1, f(0) = 0, f'(0) = 0$$
$$g(1) = 1, f(1) = 0, f'(1) = 0. \qquad (11.4)$$

We note that this problem is second order in $g$ and fourth order in $f$, and is fully determined by the six boundary conditions. Its implementation in R is:

```
swirl <- function (t, Y, eps)  {
    with(as.list(Y),
       list(c((g*f1 - f*g1)/eps,
              (-f*f3 - g*g1)/eps))
          )
    }
```

where we use the expression `with(as.list(Y),..` to make the names of `Y` available within the function.

As the derivative of `g` is second order, and the derivative of `f` is fourth order, we not only specify the initial and end values of `g` and `f` but also the first derivative of $g$, (`g1`) and the first, second and third derivatives of $f$ (`f1, f2, f3`). The first equation returned by `swirl` is of order 2, the second one of order 4. We convey this information when we solve the problem using `bvptwp(order = c(2, 4))`. When we specify the initial conditions (`yini`) we give names to each variable. Thus these names can be used in the derivative function `swirl` and it facilitates plotting.

```
eps  <- 0.001
x    <- seq(from = 0, to = 1, length = 100)
yini <- c(g = -1, g1 = NA, f = 0, f1 = 0, f2 = NA, f3 = NA)
yend <- c(1, NA, 0, 0, NA, NA)
Soltwp <- bvptwp(x = x, func = swirl, order = c(2, 4),
                 par = eps, yini = yini, yend = yend)
```

A `pairs` plot produces a pretty picture (Fig. 11.2).

```
pairs(Soltwp, main = "swirling flow III, eps=0.001")
```

Important information related to the computed numerical solution is printed by requesting the `diagnostics` of the solution. This shows the number of function

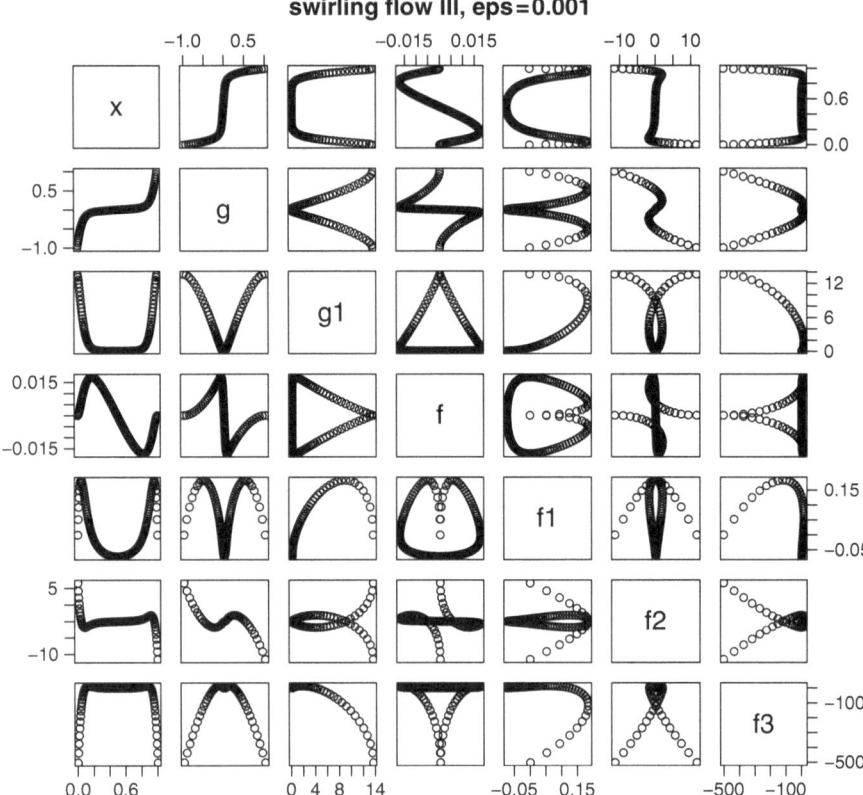

**Fig. 11.2** Pairs plot of the swirling flow III problem. See text for the R code

and boundary evaluations, the number of mesh points used as well as much more information. When the problem is solved with bvptwp, the diagnostics also provide the "conditioning parameters" which give information about the stiffness of the problem (see Sect. 10.2.4). If kappa, kappa1, gamma1, kappa2 and sigma are of moderate size, the problem is well-conditioned. If they are large, the problem is ill-conditioned. If sigma is large, the problem is stiff. If kappa1 is small and kappa, kappa2 are large the problem does not have the correct dichotomy. More information about these diagnostic parameters can be found in [10].

```
diagnostics(Soltwp)
```

```
-------------------
solved with  bvptwp
-------------------
  Integration was successful.

  1 The return code                              : 0
  2 The number of function evaluations           : 28507
```

```
 3 The number of jacobian evaluations              : 3179
 4 The number of boundary evaluations              : 84
 5 The number of boundary jacobian evaluations     : 66
 6 The number of steps                             : 18
 7 The number of mesh resets                       : 1
 8 The maximal number of mesh points               : 1000
 9 The actual number of mesh points                : 199
10 The size of the real work array                 : 280660
11 The size of the integer work array              : 14018
```

```
--------------------
conditioning pars
--------------------

 1 kappa1   :  12601.34
 2 gamma1   :  818.5373
 3 sigma    :  36.8086
 4 kappa    :  13175.8
 5 kappa2   :  574.46
```

For this problem the values of all the conditioning parameters grow when $\varepsilon$ decreases, and we could find the solution by requiring the default tolerance ($10^{-8}$).

However with the default value of atol the problem cannot be solved with very small values of eps. An error is produced if we try to do so.

```
eps <- 1e-5
Soltwp2 <- bvptwp(x = x, func = swirl, order = c(2, 4),
                  par = eps, yini = yini, yend = yend)
```

```
Error in bvpsolver(1, yini, x, func, yend, parms, order,    :
  The Expected No. Of mesh points Exceeds Storage Specifications.
```

If we use atol = 1e-4 the solution can be computed:

```
eps <- 1e-5
Soltwp2 <- bvptwp(x = x, func = swirl, order = c(2, 4),
                  par = eps, yini = yini, yend = yend, atol = 1e-4)
diagnostics(Soltwp2)
```

```
--------------------
solved with  bvptwp
--------------------

  Integration was successful.

 1 The return code                                 : 0
 2 The number of function evaluations              : 462680
 3 The number of jacobian evaluations              : 54547
 4 The number of boundary evaluations              : 576
 5 The number of boundary jacobian evaluations     : 426
 6 The number of steps                             : 107
 7 The number of mesh resets                       : 2
 8 The maximal number of mesh points               : 1000
```

```
 9 The actual number of mesh points          : 689
10 The size of the real work array           : 280660
11 The size of the integer work array        : 14018
```

```
--------------------
conditioning pars
--------------------

 1 kappa1   : 11510480
 2 gamma1   : 71573.68
 3 sigma    : 393.5928
 4 kappa    : 9.679255e+12
 5 kappa2   : 9.679252e+12
```

The conditioning parameters show that this problem is very ill-conditioned, which explains why we were not able to compute a solution with a looser relative tolerance.

## 11.4   More Complex Initial or End Conditions

The previous problem was relatively simple in the sense that the initial and final values were either a scalar or an unknown, and so they could be represented as two vectors (yini and yend). Boundary value problems can also have more complex boundary conditions, for instance if they represent relationships between several variables. As an example, we consider problem musn as described in [5].

The problem is:

$$u' = 0.5u(w - u)/v$$
$$v' = -0.5(w - u)$$
$$w' = (0.9 - 1000(w - y) - 0.5w(w - u))/z \qquad (11.5)$$
$$z' = 0.5(w - u)$$
$$y' = -100(y - w),$$

defined on the interval $[0,1]$ and subject to boundary conditions:

$$u(0) = v(0) = w(0) = 1$$
$$z(0) = -10 \qquad (11.6)$$
$$w(1) = y(1).$$

Note the last boundary condition which expresses w as a function of y at the end of the integration interval. Because of this form of the boundary conditions, yend cannot be simply input as a vector, as in the previous examples. Instead, we should write the boundary conditions as a function. Implementation of the ODE function in R is:

```
musn <- function(x, Y, pars) {
   with (as.list(Y),  {
      du <- 0.5 * u * (w - u) / v
      dv <- -0.5 * (w - u)
      dw <- (0.9 - 1000 * (w - y) - 0.5 * w * (w - u))/z
      dz <- 0.5 * (w - u)
      dy <- -100 * (y - w)
      return(list(c(du, dv, dw, dz, dy)))
   })
}
```

The boundary conditions are now implemented via a boundary function bound. For
each boundary condition (i), this function returns the residual; i.e. the boundary
condition $u(0) = 1$ is specified as $u(0) - 1$, and so on. The first four boundary
conditions are defined on the left of the integration interval. This information will
be passed as an argument to the solver (see below).

```
bound <- function(i, Y, pars) {
   with (as.list(Y),  {
      if (i == 1) return (u - 1)
      if (i == 2) return (v - 1)
      if (i == 3) return (w - 1)
      if (i == 4) return (z + 10)
      if (i == 5) return (w - y)
   })
}
```

Before proceeding with the integration, we note that this problem can only be solved
if initiated with a guess of the solution which is sufficiently close to the actual
solution. Such a guess can be provided to the solver as the vector xguess and
the matrix yguess

```
xguess <- seq(0, 1, length.out = 5)
yguess <- matrix(ncol = 5,
                 data = (rep(c(1, 1, 1, -10, 0.91), 5)))
rownames(yguess) <- c("u", "v", "w", "z", "y")
xguess
```

```
[1] 0.00 0.25 0.50 0.75 1.00
```

```
yguess
```

|   | [,1]   | [,2]   | [,3]   | [,4]   | [,5]   |
|---|--------|--------|--------|--------|--------|
| u |  1.00  |  1.00  |  1.00  |  1.00  |  1.00  |
| v |  1.00  |  1.00  |  1.00  |  1.00  |  1.00  |
| w |  1.00  |  1.00  |  1.00  |  1.00  |  1.00  |
| z | -10.00 | -10.00 | -10.00 | -10.00 | -10.00 |
| y |  0.91  |  0.91  |  0.91  |  0.91  |  0.91  |

Note that the rows of yguess have been given a name, so that this name can be
used in the derivative function (musn) and in the boundary function (bound), and
to label the output.

We also need to specify that the first four boundary conditions in function bound
are defined on the left point of the integration interval (leftbc). We increase the
precision by using a low value of atol.

```
Sol <- bvptwp(x = x, func = musn, bound = bound,
                 xguess = xguess, yguess = yguess,
                 leftbc = 4, atol = 1e-10)
```

There is more than one solution to this problem. If we start the solution with
different initial guesses, we can often obtain a second solution; we use bvpcol
here to find the second solution.

```
yguess <- matrix(ncol = 5, data = (rep(c(1,1,1, 10, 0.91), 5)))
rownames(yguess) <- c("u", "v", "w", "z", "y")
Sol2<- bvpcol(x = x, func = musn, bound = bound,
                 xguess = xguess, yguess = yguess,
                 leftbc = 4, atol = 1e-10)
```

We show only the solution curves for the variable "y" for both solutions (11.3)
(Fig. 11.3):

```
plot(Sol, Sol2, which = "y", lwd = 2)
```

## 11.5   Solving a Boundary Value Problem Using Continuation

### 11.5.1   Manual Continuation

The following non linear BVP ODE is Problem 19 from the test problems available
from [7]:

$$\xi y'' + \exp(y)y' - \frac{\pi}{2}\sin(\pi x/2)\exp(2y) = 0$$

$$y_{(x=0)} = y_{(x=1)} = 0.$$

(11.7)

This is implemented as

```
Prob19 <- function(x, y, eps) {
    pix = pi*x
    list(c(y[2],
        (pi/2*sin(pix/2)*exp(2*y[1])-exp(y[1])*y[2])/eps))
    }
```

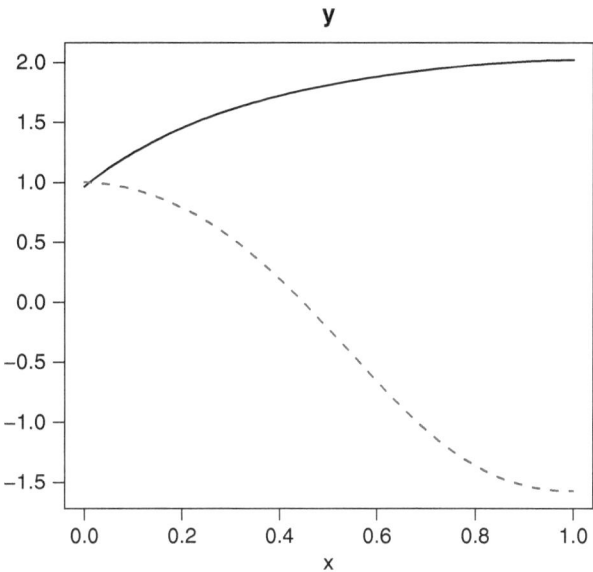

**Fig. 11.3** Two solutions of the musn problem, obtained by starting with different initial guesses. See text for the R code

The solution using bvptwp is computed without any problem when $\varepsilon = 10^{-2}$

```
x <- seq(0, 1, by = 0.01)
eps <- 1e-2
mod1 <- bvptwp(func = Prob19, yini = c(0, NA), yend = c(0, NA),
               x = x,                par = eps)
diagnostics(mod1)
```

```
--------------------
solved with  bvptwp
--------------------
  Integration was successful.

  1 The return code                            : 0
  2 The number of function evaluations         : 18057
  3 The number of jacobian evaluations         : 3091
  4 The number of boundary evaluations         : 40
  5 The number of boundary jacobian evaluations : 26
  6 The number of steps                        : 29
  7 The number of mesh resets                  : 1
  8 The maximal number of mesh points          : 1000
  9 The actual number of mesh points           : 150
 10 The size of the real work array            : 56108
 11 The size of the integer work array         : 6006
```

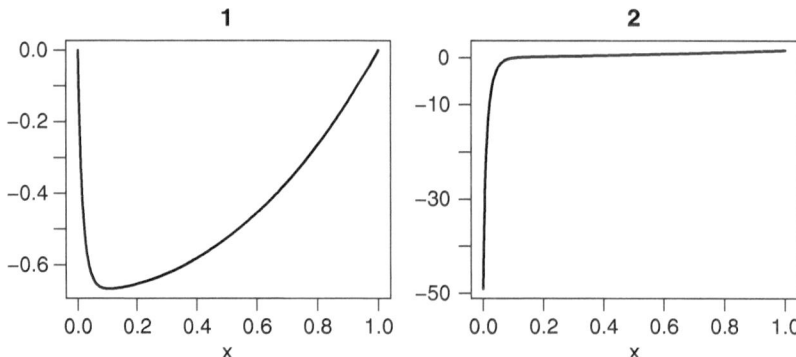

**Fig. 11.4**  Solution of the test Problem 19, for $\varepsilon = 0.01$. See text for the R code

```
--------------------
conditioning pars
--------------------

  1 kappa1   :  125.6703
  2 gamma1   :  2.236132
  3 sigma    :  93.77898
  4 kappa    :  168.431
  5 kappa2   :  42.7607
```

The problem is well-conditioned and in Fig. 11.4 we plot the solution

```
plot(mod1,   lwd = 2)
```

When we try to solve the same problem with $\varepsilon = 10^{-3}$ and with bvpcol, the code fails. However, it is possible to solve the problem if the previous solution (mod1) with eps = 0.01, is used as an initial guess for a smaller value of eps. The technique of using previous solutions and grids as starting values for the next simulations is called *continuation*.

```
xguess <- mod1[,1]
yguess <- t(mod1[,2:3])
```

```
eps <- 1e-3
mod2 <- bvpcol(func = Prob19, yini = c(0, NA), yend = c(0, NA),
               x = x, par = eps,
               xguess = xguess, yguess = yguess)
```

and we can repeat this for smaller and smaller values of eps.

## 11.5.2 Automatic Continuation

Suppose we are trying to solve a model with very small values of $\varepsilon$. We could iteratively decrease $\varepsilon$ and see how far we get with manual continuation, as illustrated in the previous section. For a more systematic approach, the functions bvptwp and bvpcol feature an *automatic continuation* strategy. The idea of automatic continuation is that the algorithm starts with a rather large value of $\varepsilon$ (argument epsini), which gives a problem that is relatively easy to solve, and lets the solver decide upon the continuation steps to perform, so that ultimately we arrive at the desired small value of $\varepsilon$ (input argument eps). The default value of epsini is 0.5, so if we are content with this value we do not need to specify it.

When the desired value of $\varepsilon$ is too small, the solver will return the solution with the smallest possible value of $\varepsilon$.

For the Problem 19, using bvpcol, this is implemented in R as:

```
eps <- 1e-7
mod2 <- bvpcol(func = Prob19, yini = c(0, NA), yend = c(0, NA),
               x = x, par = eps, eps = eps, atol = 1e-4)
diagnostics(mod2)
```

```
--------------------
solved with  bvpcol
--------------------

  Integration was successful.

 1 The return code                               : 1
 2 The number of function evaluations            : 22749
 3 The number of jacobian evaluations            : 4568
 4 The number of boundary evaluations            : 172
 5 The number of boundary jacobian evaluations   : 96
 6 The number of continuation steps              : 10
 7 The number of succesfull continuation steps   : 10
 8 The actual number of mesh points              : 50
 9 The number of collocation points per subinterval : 4
10 The number of equations                       : 2
11 The number of components (variables)          : 2

   The problem was solved for final eps equal to : 1e-07
```

If we want information about the conditioning of the problem we should use the continuation algorithm based on the Lobatto basis as implemented in bvptwp:

```
mod3 <- bvptwp(func = Prob19, yini = c(0, NA), yend = c(0, NA),
               x = x, par = eps,  eps = eps, atol=1e-4)
diagnostics(mod3)
```

```
--------------------
solved with  bvptwp
--------------------

   Integration was successful.

   1 The return code                              : 0
   2 The number of function evaluations           : 289479
   3 The number of jacobian evaluations           : 57829
   4 The number of boundary evaluations           : 740
   5 The number of boundary jacobian evaluations  : 310
   6 The number of steps                          : 32
   7 The number of mesh resets                    : 22
   8 The maximal number of mesh points            : 1000
   9 The actual number of mesh points             : 473
  10 The size of the real work array              : 54098
  11 The size of the integer work array           : 6006


--------------------
conditioning pars
--------------------

   1 kappa1  : 12397400
   2 gamma1  : 2.578484
   3 sigma   : 7627889
   4 kappa   : 16446840
   5 kappa2  : 4049448

   The problem was solved for final eps equal to : 1e-07
```

We note that the conditioning of the problems grows like $1/\varepsilon$.


## 11.6   BVPs with Unknown Constants


Not all boundary value problems are defined in the form required to be solved by functions bvptwp or bvpcol. However, we can often use tricks to convert them into the desired form [2].

For instance, a wide range of boundary value problems involve unknown constants, the value of which should be part of the solution. Such problems are not defined in the form required to be solved by functions bvptwp or bvpcol. We can circumvent this by treating the unknown constant as a variable of which the first derivative is zero, and we add the corresponding equation to the system of differential equations. In the following examples we show that this trick can be also used to deal with periodic boundary conditions and with unknown integration intervals.

### 11.6.1   The Elastica Problem

We now implement the *elastica* problem.[1] The problem consists of a system of
five differential equations, describing an elastica in the $(x,y)$ plane. The angle $\phi$ is
modeled rather than the separate derivatives for the $x$ and $y$ components. This gives
control over the curvature $\kappa$.

The original system reads:

$$x' = \cos(\phi)$$
$$y' = \sin(\phi)$$
$$\phi' = \kappa \tag{11.8}$$
$$\kappa' = F\cos(\phi),$$

where $F$ is an (unknown) constant. The problem has five boundary conditions:

$$x(0) = 0$$
$$y(0) = 0$$
$$\kappa(0) = 0 \tag{11.9}$$
$$y(0.5) = 0$$
$$\phi(0.5) = -\pi/2.$$

These five boundary conditions make the problem fully specified. Indeed, apart from
the four integration constants arising from the four first order differential equations,
the unknown constant $F$ should also be determined. To solve this problem with
bvpcol we add to the system (11.8) the equation for the parameter $F$:

$$F' = 0 \tag{11.10}$$

The full problem then consists of the ODE (11.8) and (11.10), with the boundary
conditions given by (11.9). The implementation in R

```
Elastica <- function (x, y, pars) {
   list( c(cos(y[3]),
           sin(y[3]),
           y[4],
           y[5] * cos(y[3]),
           0))
}
bvpsol <- bvpcol(func = Elastica,
         yini = c(x = 0,  y = 0, p = NA,    k = 0,  F = NA),
         yend = c(x = NA, y = 0, p = -pi/2, k = NA, F = NA),
         x = seq(from = 0, to = 0.5, by = 0.01))
```

[1] From (http://www.ma.ic.ac.uk/~jcash/BVP_software).

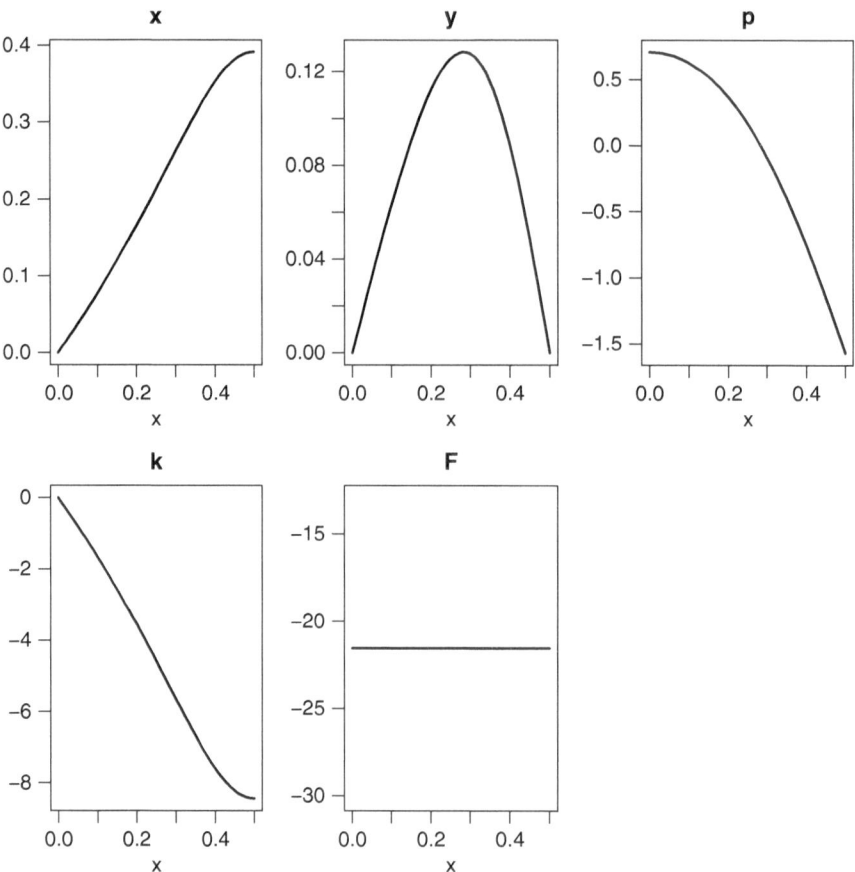

**Fig. 11.5** Solution of the elastica problem. See text for the R code

```
plot(bvpsol, lwd = 2)
```

Note how *F* remains constant (Fig. 11.5); its value is:

```
bvpsol[1,"F"]
```

```
        F
-21.54909
```

## 11.6.2  *Non-separated Boundary Conditions*

In many cases of practical interest, each boundary condition is defined either at one or at the other end of the integration interval; so-called *separated* boundary

**Fig. 11.6** Schematic diagram
of the measel problem

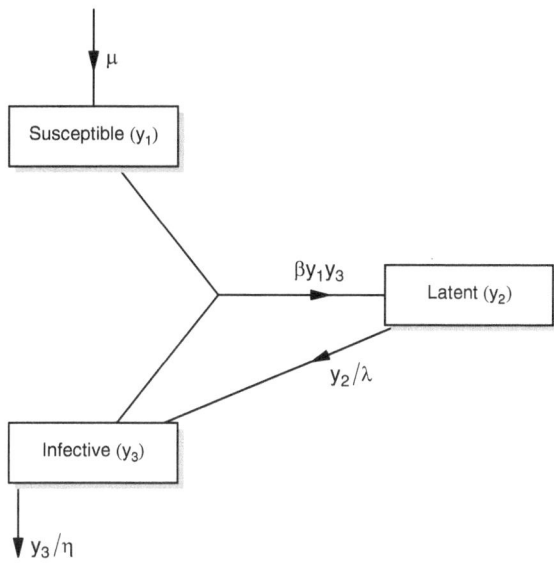

conditions. A special case of *non-separated* boundary conditions are periodic boundary conditions, which specify that variables have the same value at both ends of the integration interval.

Functions bvptwp and bvpcol accept only problems with separated boundary conditions. To use them also for solving boundary value problems with periodic boundary conditions, the unknown boundary condition is considered a "parameter" which has to be estimated. Thus the equations are augmented with dependent variables whose derivatives are 0, one for each boundary condition.

Consider as an example the problem *measels* from [2]. This problem describes the outbreak of the disease by considering a given population that consists of four categories, susceptible (S), infectives (I), latents (L) and immunes (M).

Assuming that the population density is constant (N), we have:

$$S + I + L + M = N. \tag{11.11}$$

The model describes the relative proportion of susceptibles ($y_1 = S/N$), latents ($y_2 = L/N$) and infected individuals ($y_3 = I/N$).

The dynamics of the disease are expressed as (see Fig. 11.6):

$$\begin{aligned}
y_1' &= \mu - \beta y_1 y_3 \\
y_2' &= \beta y_1 y_3 - y_2/\lambda \\
y_3' &= y_2/\lambda - y_3/\eta,
\end{aligned} \tag{11.12}$$

on the interval [0, 1]. To add seasonality to this model, the infection rate $\beta$ is described by a cosine function, i.e. $\beta = \beta_0(1 + \cos(2\pi t))$, where $\beta_0 = 1575$. The other parameters are $\mu = 0.02, \lambda = 0.0279, \eta = 0.1$.

Now it is assumed that the outbreak of measels has a recurrent pattern, i.e. the solution is periodic, that is $y_i(0) = y_i(1)$ for all $i$.

We translate this problem into a two-point separated boundary value problem by introducing three additional unknowns, $y_4, y_5, y_6$ that represent the boundary conditions. For instance, $y_4$ is the boundary value of $y_1$, i.e. $y_4 = y_1(0) = y_1(1)$ and so on. Being "constants", we specify their derivatives to be = 0. The derivative and boundary function are, respectively [2]:

```
measel <- function(t, y, pars)  {
    bet <- 1575 * (1 + cos(2 * pi * t))
    dy1 <- mu - bet * y[1] * y[3]
    dy2 <- bet * y[1] * y[3] - y[2] / lam
    dy3 <- y[2] / lam - y[3] / eta
    dy4 <- 0
    dy5 <- 0
    dy6 <- 0
    list(c(dy1, dy2, dy3, dy4, dy5, dy6))
}
bound <- function(i, y, pars) {
    if ( i == 1 | i == 4) return( y[1] - y[4])
    if ( i == 2 | i == 5) return( y[2] - y[5])
    if ( i == 3 | i == 6) return( y[3] - y[6])
}
```

The model is solved by first computing a solution using the shooting method. Shooting fails if the initial guesses are all 0, (the default), so we take all initial guesses to be 1 instead. Also, we impose very loose tolerances.

```
mu   <- 0.02 ; lam <- 0.0279 ; eta <- 0.1
x <- seq(from = 0, to = 1, by = 0.01)
Sola <- bvpshoot(func = measel, bound = bound,
        x = x, leftbc = 3, atol = 1e-12, rtol = 1e-12,
        guess = c(y1 = 1, y2 = 1, y3 = 1, y4 = 1, y5 = 1, y6 = 1))
```

We also need to provide a reasonable initial guess in order to have a solution using bvptwp. We use the same initial guess as for the shooting method

```
yguess <- matrix(ncol = length(x), nrow = 6, data = 1)
rownames(yguess) <- paste("y", 1:6, sep="")
Sol  <- bvptwp  (func = measel, bound = bound,
          x = x, leftbc = 3, xguess = x, yguess = yguess)
```

the boundaries are indeed periodic

```
max(abs(Sol[1,-1]  - Sol[nrow(Sol),-1]))
```

[1] 0

---

[2] The '—' in function bound is the 'OR' operator.

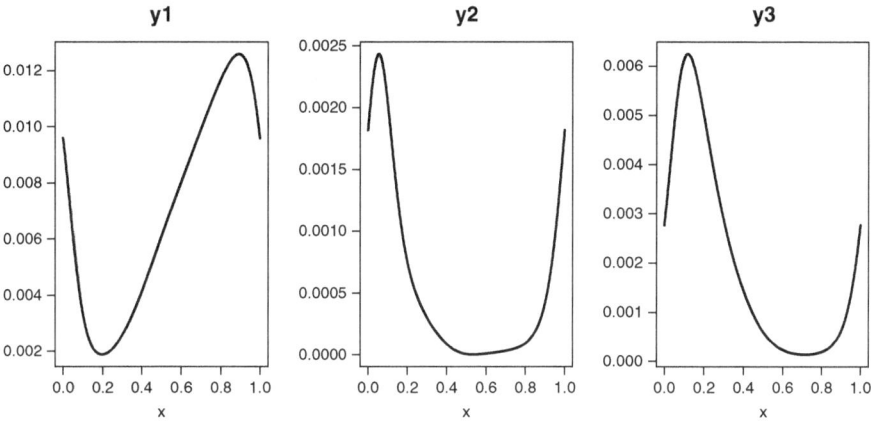

**Fig. 11.7** Solution of the measel problem, comprising periodic boundary conditions. See text for the R code

```
max(abs(Sola[1,-1]  -  Sola[nrow(Sola),-1]))
```

```
[1]  1.130485e-13
```

We plot the number of susceptible, latent and infective individuals (the first three variables), arranged in one row, three columns (mfrow):

```
plot(Sol, lwd = 2, which = 1:3, mfrow = c(1, 3))
```

### 11.6.3   An Unknown Integration Interval

In some problems, the unknown constant specifies the size of the integration interval. In this case, a change of independent variable $(x)$ is carried out, i.e. $x$ is replaced by $\tau = x/b$ where $b$ is the (unknown) integration interval (see Sect. 10.3.3). Now the ODE system, in the new independent variable becomes (Fig. 11.7):

$$\frac{dy}{d\tau} = bf(b\tau, y)$$
$$\frac{db}{d\tau} = 0,$$

(11.13)

where $\tau$ is between 0 and 1.

A BVP that combines cyclic boundary conditions with an unknown integration interval is the nerve impulse model, a problem described in [14] and treated in [16]. This model is given by

$$y_1' = 3(y_1 + y_2 - 1/3y_1^3 - 1.3)$$
$$y_2' = -(y_1 - 0.7 + 0.8y_2)/3.$$

(11.14)

The problem is defined on the interval $[0, T]$, where $T$ is unknown and subject to boundary conditions:

$$y_1(0) = y_1(T)$$
$$y_2(0) = y_2(T).$$

(11.15)

Based on (11.13) the original equations (11.14) are rewritten as:

$$y_1' = 3T(y_1 + y_2 - 1/3y_1^3 - 1.3)$$
$$y_2' = -T(y_1 - 0.7 + 0.8y_2)/3$$
$$T' = 0,$$

(11.16)

defined on the interval $[0, 1]$ and with boundary conditions:

$$y_1(0) = y_1(1)$$
$$y_2(0) = y_2(1).$$

(11.17)

An extra boundary condition is needed to determine the unknown parameter $T$. The model described in [16] uses the condition that $dy_2/d\tau(0) = 1$:

$$1 = -T(y_1(0) - 0.7 + 0.8y_2(0))/3.$$

(11.18)

We again use the strategy of defining the parameters as extra dependent variables, with derivative = 0. The augmented derivative function is, with variable $y_3$ the unknown parameter $T$, and the variables $y_4$ and $y_5$ representing the initial conditions for $y_1$, and $y_2$ respectively:

```
nerve <- function (x, y, p)
    list(c(3 * y[3] * (y[1] + y[2] - 1/3 * (y[1]^3) -1.3),
           (-1/3) * y[3] * (y[1] - 0.7 + 0.8 * y[2]) ,
           0,
           0,
           0)
    )
```

The required boundary function, with the first three boundary conditions imposed at the left boundary is:

```
bound <- function(i, y, p) {
    if (i ==1) return (-y[3]* (y[1] - 0.7 + 0.8*y[2])/3 -1)
    if (i ==2) return (y[1] - y[4] )
    if (i ==3) return (y[2] - y[5] )
    if (i ==4) return (y[1] - y[4] )
```

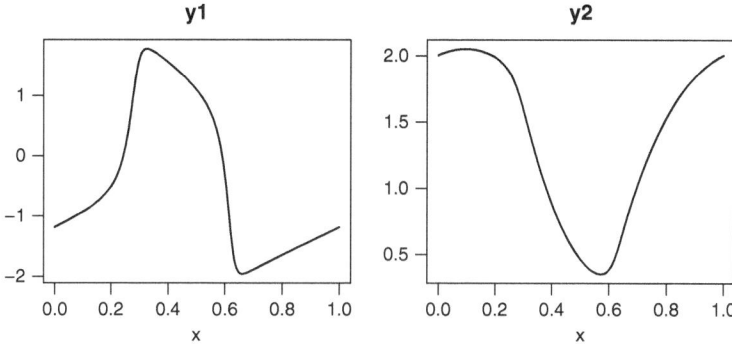

**Fig. 11.8** Solution of the nerve impulse problem, comprising two state variables and three parameters. See text for the R code

```
    if (i ==5) return (y[2] - y[5] )
  }
```

Boundary condition 2 sets the left boundary of $y_1$ equal to parameter $y_4$, boundary condition 4 does the same for the right boundary of $y_1$.

In order to solve this problem, we also require good approximations to the variables. Since we are looking for periodic boundary conditions, we take a sine and cosine for the first two state variables; the constants obtain the value 5 [16] (Fig. 11.8):

```
xguess <- seq(0, 1, by = 0.1)
yguess <- matrix(nrow = 5, ncol = length(xguess), data = 5.)
yguess[1,] <- sin(2 * pi * xguess)
yguess[2,] <- cos(2 * pi * xguess)
rownames(yguess) <- c("y1", "y2", "T", "y1ini", "y2ini")
```

```
Sol   <- bvptwp(func = nerve, bound = bound,
                x = seq(0, 1, by = 0.01),leftbc = 3,
                xguess = xguess, yguess = yguess)
```

The first row of Sol shows the initial conditions, and the values of the constant parameters T, y1ini, y2ini.

```
Sol[1,]
```

```
       x          y1          y2           T      y1ini       y2ini
0.000000 -1.183453    2.004203 10.710808  -1.183453    2.004203
```

```
plot(Sol, lwd = 2, which = c("y1", "y2"))
```

## 11.7   Integral Constraints

Another common form of a two-point boundary value problem is when an integral constraint is imposed. Such a constraint typically has the form:

$$\int_a^b G(t, y(t))dt = \alpha. \tag{11.19}$$

If we now define

$$I(x) = \int_a^x G(t, y(t))dt, \tag{11.20}$$

we can replace the integral constraint by

$$\begin{aligned} I'(x) &= G(x, y(x)) \\ I(a) &= 0 \\ I(b) &= \alpha. \end{aligned} \tag{11.21}$$

Consider for example the following problem

$$\begin{aligned} u_1' &= u_2 \\ u_2' &= -p\exp(u_1) \\ u_1(0) &= 1 \\ \int_0^1 u_2 dx &= 1. \end{aligned} \tag{11.22}$$

This is rewritten as:

$$\begin{aligned} u_1' &= u_2 \\ u_2' &= -p\exp(u_1) \\ I' &= u_2 \\ u_1(0) &= 1 \\ I(0) &= 0 \\ I(1) &= 1, \end{aligned} \tag{11.23}$$

and implemented in R as:

```
library(bvpSolve)
integro <- function (t, u, p)
   list(c(u[2], -p*exp(u[1]), u[2]))
yini <- c(u1 = 1, u2 = NA, I = 0)
yend <- c(NA, NA, 1)
x    <- seq(from = 0, to = 1, by = 0.01)
out <- bvpcol (yini = yini, yend = yend, func = integro,
                x = x, parms = 0.5)
```

An important special case of such an integral constraint arises in the computation of homoclinic orbits [13]. In this application the norm of the solution is specified. Setting $G(x, y(x)) = y^T(x)y(x)$ it follows that $y(b) = ||y_2||^2$.

Similar tricks can be used to convert integro-differential equations and some optimal control problems to standard form [5, p. 473].

## 11.8   Sturm-Liouville Problems

Another very important class of boundary value problems that can be converted to standard form are eigenvalue problems. Of particular interest are Sturm-Liouville problems [5, p. 478], which are defined as

$$
\begin{aligned}
-(py')' + qy &= \lambda ry \quad a \le x \le b \\
C_1 y(a) + p(a)y'(a) &= 0 \\
C_2 y(b) - p(b)y'(b) &= 0,
\end{aligned}
\tag{11.24}
$$

where $p$, $q$, and $r$ are continuous functions with $p > 0$, $q > 0$, $r > 0$ and with $-\infty < a < b < \infty$, and $C_1$, $C_2$ both greater than zero. In the linear case we have a homogeneous ODE with a homogeneous boundary condition with one or more of them depending on an unknown parameter.

Perhaps the most simple example of this type of eigenvalue problem is:

$$
\begin{aligned}
y'' + \lambda y &= 0 \\
y(0) &= 0 \\
y(\pi) &= 0,
\end{aligned}
\tag{11.25}
$$

where we have chosen separated boundary conditions. For all values of $\lambda$, the function $y(x) = 0$ is a solution of this eigenvalue problem. However, the aim is to find values of the parameter $\lambda$ so that the differential equation also has non-trivial solutions. These values of the parameter are called the eigenvalues, and the corresponding non-trivial solutions are the eigenvectors. It is easy to see that, if $y(x)$ is a solution of (11.25), then so is $ay(x)$ for any constant $a$. This means that we need to prescribe a normalising condition to specify exactly which solution we are interested in. This normalising condition can be regarded as an extra boundary condition, so that we now have three such conditions. The first two boundary conditions are required so that we can compute the general solution, while the third boundary condition allows us to pick precisely which solution we require. The question of which normalising condition to pick for use with (11.25) is addressed in [5] but we note that often a reasonable choice is $y'(0) = 1$.

We can now implement the eigenvalue problem. First we rewrite it in first order form, and add the fact that $\lambda$ is a constant, hence its derivative is 0:

$$y'_1 = y_2$$
$$y'_2 = -\lambda y_1$$
$$\lambda' = 0$$
$$y(0) = 0$$
$$y(\pi) = 0$$
$$y'(0) = 1.$$

(11.26)

A very appealing approach to the solution of this problem is to use simple shooting.

```
Sturm <- function(x, y, p) {
    dy1 <- y[2]
    dy2 <- -y[3] * y[1]
    dy3 <- 0.
    list( c(dy1, dy2, dy3))
 }
yini <- c(y = 0, dy = 1,  lambda = NA)
yend <- c(y = 0, dy = NA, lambda = NA)
x    <- seq(from = 0, to = pi,  by = pi/10)
S1 <- bvpshoot(yini = yini, yend = yend, func = Sturm,
                parms = 0, x = x)
```

The eigenvalue of the problem is equal to 1:

```
(lambda1 <- S1[1, "lambda"])
```

```
lambda
    1
```

The analytic solution of this problem is known [15], and can be compared with the numerical solution produced using shooting (S1).

```
ana <- function(x, lambda) sin(x*sqrt(lambda))/sqrt(lambda)
max (abs(S1[,2]-ana(S1[,1],lambda1)))
```

```
[1] 8.987562e-08
```

## 11.9    A Reaction Transport Problem

A very simple and straightforward way of solving BVPs that fall into the category of reaction transport problems, is to use the strategy as described in the PDE (Chap. 8, Sect. 8.3). In this technique the differential equations are represented by finite difference functions from the R package **ReacTran** [18] (Chap. 9) and a suitable solver from the R package **rootSolve** [17] is used to obtain the solution.

To illustrate this, we implement a biogeochemical model that describes nitrate ($HNO_3$), ammonia ($NH_3$) and oxygen ($O_2$) in a highly polluted estuary. The main biogeochemical process is the nitrification, which is the oxidation of ammonia to nitrate, consuming oxygen:

$$NH_3 + 2O_2 \rightarrow HNO_3 + H_2O \tag{11.27}$$

Nitrate, ammonia and oxygen enter the estuary with the flow upstream, and are transported along the spatial axis; tidal mixing is represented as dispersion. Assuming that the concentrations are at steady state (time derivatives $= 0$), the mass balances for ammonia, nitrate and oxygen are:

$$
\begin{aligned}
0 &= \frac{\partial}{\partial x}\left(D\frac{\partial HNO_3}{\partial x}\right) - v\frac{\partial HNO_3}{\partial x} + r_{nit} \\
0 &= \frac{\partial}{\partial x}\left(D\frac{\partial NH_3}{\partial x}\right) - v\frac{\partial NH_3}{\partial x} - r_{nit} \\
0 &= \frac{\partial}{\partial x}\left(D\frac{\partial O_2}{\partial x}\right) - v\frac{\partial O_2}{\partial x} - 2r_{nit} + p(O_2s - O_2) \\
r_{nit} &= rNH_3\frac{O_2}{O_2 + k},
\end{aligned}
\tag{11.28}
$$

where $r_{nit}$ is the nitrification (parameters $k = 1$, $r = 0.1$) and the last equation in the mass balance of oxygen is reaeration ($p = 0.1$, $O_2s = 300$), which adds oxygen to the water. Flow velocity and the dispersion coefficient are constant ($v = 1,000$, $D = 10^7$). The entire estuary is 100,000 m long, and has constant cross-sectional surface. The boundary conditions are:

$$
\begin{aligned}
NH_3(0) &= 500, O_2(0) = 50, HNO_3(0) = 100 \\
NH_3(1e^5) &= 10, O_2(1e^5) = 30, HNO_3(1e^5) = 250.
\end{aligned}
\tag{11.29}
$$

Before we discuss the implementation in R, you may want to refresh your knowledge on how to use the functions of the R packages **ReacTran** and **rootSolve** (see Chap. 9).

We start by loading the R package **ReacTran**, define the one-dimensional grid (there are 1,000 grid cells), and the parameters. In the derivative function (Estuary) the dependent variable vector (y) is first split into the three described species, after which their advective-diffusive transport is performed using **ReacTran**'s function tran.1D. Next the rates (reaeration, r_nit) are calculated and the derivatives returned, combined as a vector (c()) and packed as a list.

This problem is solved using **rootSolve**'s function steady.1D, which implements a Newton-Raphson method. As initial guess for the root we simply provide 3*N random numbers (runif); we need to specify the dimensionality of the problem (dimens). As none of the species can become negative, we force the root finding algorithm to retrieve only positive values. We return the time it takes to solve these 3,000 equations.

```
library(ReacTran)
N      <- 1000
Grid <- setup.grid.1D(N = N, L = 100000)
v <- 1000; D <- 1e7; O2s <- 300;
NH3in <- 500; O2in <- 100; NO3in <- 50
r <- 0.1; k <- 1.; p <- 0.1
Estuary <- function(t, y, parms)   {
  NH3 <- y[1:N]
  NO3 <- y[(N+1):(2*N)]
  O2  <- y[(2*N+1):(3*N)]
  tranNH3<- tran.1D (C = NH3, D = D, v = v,
            C.up = NH3in, C.down = 10,  dx = Grid)$dC
  tranNO3<- tran.1D (C = NO3, D = D, v = v,
            C.up = NO3in,  C.down = 30,  dx = Grid)$dC
  tranO2 <- tran.1D (C = O2 , D = D, v = v,
            C.up = O2in, C.down = 250, dx = Grid)$dC

  reaeration <- p * (O2s - O2)
  r_nit        <- r * O2 / (O2 + k) * NH3

  dNH3    <- tranNH3 - r_nit
  dNO3    <- tranNO3 + r_nit
  dO2     <- tranO2  - 2 * r_nit + reaeration

  list(c(  dNH3, dNO3, dO2 ))
  }
print(system.time(
  std <- steady.1D(y = runif(3 * N), parms = NULL,
            names=c("NH3", "NO3", "O2"),
            func = Estuary, dimens = N,
            positive = TRUE)
))
```

```
user    system elapsed
0.17     0.00     0.17
```

It is instructive to solve the problem with a reduced concentration of ammonia in the inflowing water (NH3in). Here we take the ammonia concentration equal to 100.

```
NH3in <- 100
std2 <- steady.1D(y = runif(3 * N), parms = NULL,
          names=c("NH3", "NO3", "O2"),
          func = Estuary, dimens = N,
          positive = TRUE)
```

We plot how the three species change along the estuary (Fig. 11.9). Note that, in the case where ammonia concentrations upstream are high, oxygen is completely exhausted in the upstream parts of the estuary. Unfortunately this was the case for the Belgian Scheldt until the 1990 due to input of untreated municipal waste

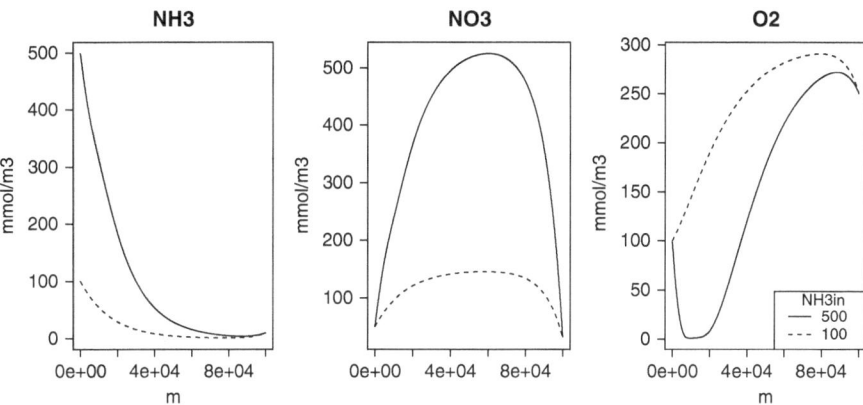

**Fig. 11.9** Solution of the estuarine problem. See text for the R code

water containing huge concentrations of ammonia [19]. The current situation in the estuary is more similar to the second scenario where ammonia concentrations in the inflowing water are much lower and the oxygen has returned (Fig. 11.9, dashed line).

```
plot(std, std2, grid = Grid$x.mid, ylab = "mmol/m3",
        xlab = "m", mfrow = c(1,3), col = "black")
legend("bottomright", lty = 1:2, title = "NH3in",
            legend = c(500, 100))
```

## 11.10 Exercises

### 11.10.1 A Stiff Boundary Value Problem

The first exercise is to implement a simple BVP ODE (which is Problem 14 from the test problems available from [7]):

$$\varepsilon y'' = y_1 - (\varepsilon \pi^2 + 1)\cos(\pi x)$$
$$y(-1) = 0 \qquad\qquad (11.30)$$
$$y(1) = 0.$$

Solve the problem for the following values of $\varepsilon$: $0.01, 0.0025, 0.0001$ (Fig. 11.10). Compare your results with the analytic solution, which is: $y(x) = \cos(\pi x) + \exp((x - 1)/\sqrt{(\varepsilon)}) + \exp(-(x + 1)/\sqrt{(\varepsilon)})$.

**Fig. 11.10** Problem 14

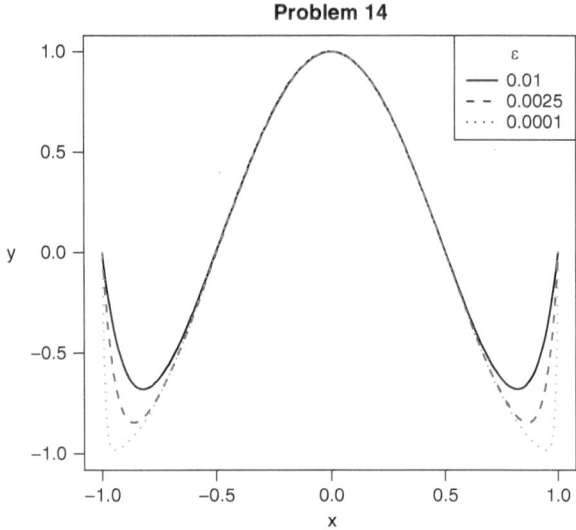

## 11.10.2   The Mathieu Equation

In the following Sturm-Liouville boundary value problem [16], the fourth eigen-value of Mathieu's equation (parameter $\lambda$) is computed. The equation is

$$y'' + (\lambda - 10\cos(2t))y = 0, \tag{11.31}$$

defined on $[0,\pi]$, and with boundary conditions

$$\begin{aligned} y'(0) &= 0 \\ y'(\pi) &= 0 \\ y(0) &= 1. \end{aligned} \tag{11.32}$$

Here all the initial values (at $t = 0$) are prescribed, in addition to one condition at the end of the interval. As $\lambda$ is unknown the problem is fully determined.

Implement this problem in R; one possible solution is in Fig. 11.11.

## 11.10.3   Another Swirling Flow Problem

Problem 1.4 of [5] is another formalisation of the swirling flow problem from Sect. 1.1.3.

**Fig. 11.11** The Mathieu
problem

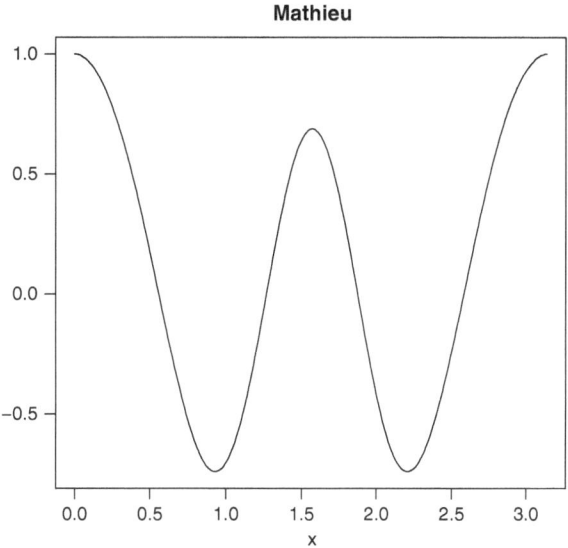

$$0 = f'''' - R[f'^2 - ff''] + RA$$
$$0 = h'' + Rfh' + 1 \qquad\qquad (11.33)$$
$$0 = \theta'' + Pef\theta' = 0,$$

and where $A$ is an (unknown) parameter and $Pe = 0.7R$ is the Peclet number; $R$ the
Reynolds number. The problem is defined on the interval $[0,1]$ and subject to eight
boundary conditions:

$$f(0) = f'(0) = 0$$
$$f(1) = f'(1) = 1$$
$$h(0) = h(1) = 0 \qquad\qquad (11.34)$$
$$\theta(0) = 0$$
$$\theta(1) = 1.$$

To solve this problem, you will need to first rewrite this system of equations as a
system of seven first order equations, and add an extra equation for the constant
parameter $A$. This problem is not simple to solve; function bvpcol is the most
efficient numerical method. Solve the equations for $R = 100$, 1,000 and 10,000. Plot
all solutions in one figure (see Fig. 11.12). What is the value of the constant in these
three cases?

**Fig. 11.12** Another swirling
flow problem

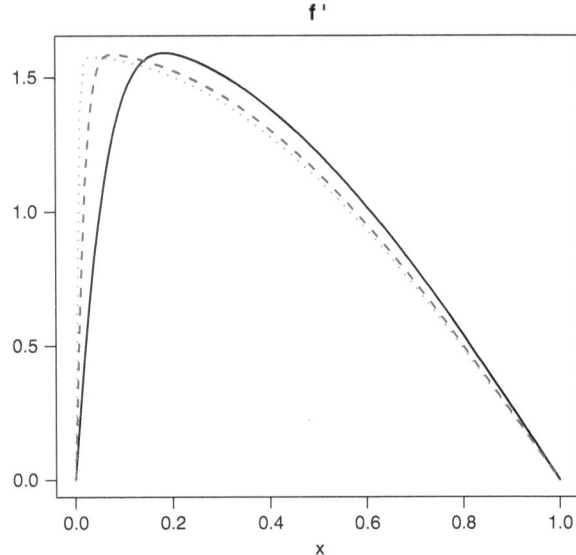

## 11.10.4 *Another Reaction Transport Problem*

Solve the following coupled second order equations using either bvpcol or bvptwp, in the interval [0,2]:

$$h'' = 100(1 - 0.1h)h$$
$$f'' = 10(1 - 0.1h)f$$
$$h(0) = f(0) = 1$$
$$h(2) = f(2) = 1.$$
(11.35)

This is a slightly simplified problem from [1], as given in [15].

If we rewrite the problem (11.35) as:

$$0 = D\frac{d^2h}{dx^2} - 100(1 - 0.1h)h$$

$$0 = D\frac{d^2f}{dx^2} - 10(1 - 0.1h)f$$
(11.36)

$$h(0) = f(0) = 1$$
$$h(2) = f(2) = 1,$$

it is clear that these equations in fact describe diffusion and reaction in one dimension with $D = 1$. Consequently another way to solve these equations is to

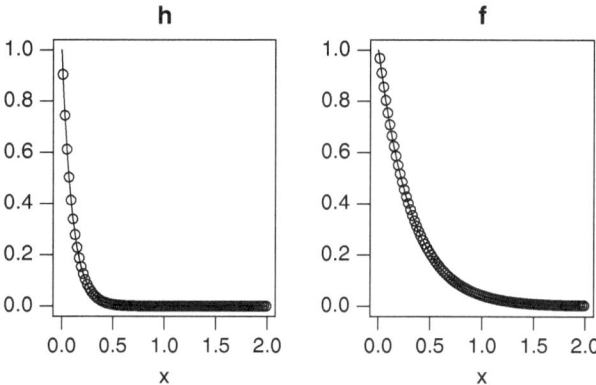

**Fig. 11.13** The transport and reaction problem, solved using `bvpcol` (*solid lines*) and `steady.1D` (points)

make use of the grid creation and finite differences offered by the R package **ReacTran** and to use function `steady.1D` from **rootSolve** to find a solution, much in the same way as we solved the problem from Sect. 11.9.

Although both methods differ considerably in the way the problem is implemented, they give the same solution (Fig. 11.13).

# References

1. Ames, W., & Lohner, E. (1981). Nonlinear models of reaction–diffusion in rivers. In R. Vichnevetsky & R. Stepleman (Eds.), *Advances in computer methods for partial differential equations* (Vol. IV, pp. 217–219). New Brunswick: IMACS.
2. Ascher, U. M., & Petzold, L. R. (1998). *Computer methods for ordinary differential equations and differential-algebraic equations.* Philadelphia: SIAM.
3. Ascher, U. M., Christiansen, J., & Russell, R. D. (1979). **COLSYS**–a collocation code for boundary value problems. In B. Childs et al. (Ed.), *Lecture notes in computer science 76* (pp. 164–185). New York: Springer.
4. Ascher, U. M., Christiansen, J., & Russell, R. D. (1981). Collocation software for boundary-value ODEs. *ACM Transactions on Mathematical Software, 7,* 209–222.
5. Ascher, U. M., Mattheij, R. M. M., & Russell, R. D. (1995). *Numerical solution of boundary value problems for ordinary differential equations.* Philadelphia: SIAM.
6. Bader, G., & Ascher, U. M. (1987). A new basis implementation for a mixed order boundary value ODE solver. *SIAM Journal on Scientific and Statistical Computing, 8,* 483–500.
7. Cash, J. R. (2007). Algorithms for the solution of two-point boundary value problems (http://www.ma.ic.ac.uk/~jcash/BVPsoftware).
8. Cash, J. R., & Mazzia, F. (2005). A new mesh selection algorithm, based on conditioning, for two-point boundary value codes. *Journal of Computational and Applied Mathematics, 184,* 362–381.
9. Cash, J. R., & Mazzia, F. (2006). Hybrid mesh selection algorithms based on conditioning for two-point boundary value problems. *Journal of Numerical Analysis, Industrial and Applied Mathematics, 1*(1), 81–90.

10. Cash, J. R., & Mazzia, F. (2009).  Conditioning and hybrid mesh selection algorithms for two-point boundary value problems. *Scalable Computing: Practice and Experience, 10*(4), 347–361.
11. Cash, J. R., & Wright, M. H. (1991).  A deferred correction method for nonlinear two-point boundary value problems: Implementation and numerical evaluation. *SIAM Journal on Scientific and Statistical Computing, 12*, 971–989.
12. Cash, J. R., Moore, G. & Wright, R. W. (1995).  An automatic continuation strategy for the solution of singularly perturbed linear two-point boundary value problems. *Journal of Computational Physics, 122*, 266–279.
13. Moore, G. (1995). Computation and parameterisation of periodic and connecting orbits. *IMA Journal of Numerical Analysis, 15*, 245–263.
14. Seydel, R. (1988). *From equilibrium to Chaos.* New York: Elsevier.
15. Shampine, L. F., Gladwell, I., & Thompson, S. (2003).  *Solving ODEs with MATLAB.* Cambridge: Cambridge University Press.
16. Shampine, L. F., Kierzenka, J., & Reichelt, M. W. (2000). *Solving boundary value problems for ordinary differential equations in MATLAB with bvp4c.* In Matlab Guide, D.J. Higham and N.J. Higham, pp 163–169, Philadelphia: SIAM.
17. Soetaert, K. (2011). **rootSolve**: *Nonlinear root finding, equilibrium and steady-state analysis of ordinary differential equations.* R package version 1.6.2.
18. Soetaert, K., & Meysman, F. (2012).  Reactive transport in aquatic ecosystems: Rapid model prototyping in the open source software R. *Environmental Modelling and Software, 32*, 49–60.
19. Soetaert, K., Middelburg, J. J., Heip, C., Meire, P., Van Damme, S., & Maris, T. (2006). Long-term change in dissolved inorganic nutrients in the heterotrophic scheldt estuary (Belgium, the Netherlands). *Limnology and Oceanography, 51*, 409–423.
20. Soetaert, K., Cash, J. R., & Mazzia, F. (2011). **bvpSolve**: *Solvers for boundary value problems of ordinary differential equations.* R package version 1.2.2.

# Appendix A

## A.1 Butcher Tableaux for Some Runge-Kutta Methods

The Butcher tableaux of the explicit Cash-Karp and the implicit Radau Runge-Kutta methods are reported in Tables A.1 and A.2.

**Table A.1** The Cash-Karp Formula. The first row of b coefficients gives the fifth order solution, while the second row gives the solution of order 4. The Cash-Karp formula is described in [7]

| 0 | 0 | | | | | |
|------|----------------|--------------|----------------|----------------|-------------|-----------|
| 1/5  | 1/5            | 0            | | | | |
| 3/10 | 3/40           | 9/40         | 0              | | | |
| 3/5  | 3/10           | −9/10        | 6/5            | 0              | | |
| 1    | −11/54         | 5/2          | −70/27         | 35/27          | 0           | |
| 7/8  | 1,631/55,296   | 175/512      | 575/13,824     | 44,275/110,592 | 253/4,096   | 0         |
|      | 37/378         | 0            | 250/621        | 125/594        | 0           | 512/1,771 |
|      | 2,825/27,648   | 0            | 18,575/48,384  | 13,525/55,296  | 277/14,336  | 1/4       |

**Table A.2** The RADAU formula of order 5

| $(4-\sqrt{6})/10$ | $(88-7\sqrt{6})/360$ | $(296-169\sqrt{6})/1,800$ | $(-2+3\sqrt{6})/225$ |
|-------------------|----------------------|---------------------------|----------------------|
| $(4+\sqrt{6})/10$ | $(296+169\sqrt{6})/1,800$ | $(88+7\sqrt{6})/360$  | $(-2-3\sqrt{6})/225$ |
| 1                 | $(16-\sqrt{6})/36$   | $(16+\sqrt{6})/36$        | 1/9                  |
|                   | $(16-\sqrt{6})/36$   | $(16+\sqrt{6})/36$        | 1/9                  |

## A.2 Coefficients for Some Linear Multistep Formulae

The coefficients for the Adams, BDF, and MEBDF formulae can conveniently be put in a table. In what follows we list the first few methods of each (Tables A.3–A.6).

K. Soetaert et al., *Solving Differential Equations in R*, Use R!,
DOI 10.1007/978-3-642-28070-2, © Springer-Verlag Berlin Heidelberg 2012

**Table A.3** Coefficients $\beta_i$ of the first few Adams-Bashforth methods $y_{n+1} = y_n + h\sum_{j=0}^{k-j}$ $\beta_j f(x_{n+1-j}, y_{n+1-j})$. The first method is the explicit Euler method. The $\alpha$ coefficients are always $\alpha_0 = -1, \alpha_1 = 1$ and are not given

| Steps $k$ | Order $p$ | $\beta_0$ | $\beta_1$ | $\beta_2$ | $\beta_3$ | $\beta_4$ | $\beta_5$ |
|---|---|---|---|---|---|---|---|
| 1 | 1 | $1$ | $0$ | | | | |
| 2 | 2 | $-\dfrac{1}{2}$ | $\dfrac{3}{2}$ | $0$ | | | |
| 3 | 3 | $\dfrac{5}{12}$ | $-\dfrac{16}{12}$ | $\dfrac{23}{12}$ | $0$ | | |
| 4 | 4 | $-\dfrac{9}{24}$ | $\dfrac{37}{24}$ | $-\dfrac{59}{24}$ | $\dfrac{55}{24}$ | $0$ | |
| 5 | 5 | $\dfrac{251}{720}$ | $-\dfrac{1,274}{720}$ | $\dfrac{2,616}{720}$ | $-\dfrac{2,774}{720}$ | $\dfrac{1,901}{720}$ | $0$ |

**Table A.4** Coefficients $\beta_i$ of the first few Adams-Moulton formulas. $y_{n+1} = y_n + h\sum_{j=0}^{k}$ $\beta_{k-j} f(x_{n+1-j}, y_{n+1-j})$. The first two Adams-Moulton formulae are the implicit Euler and the trapezium method respectively. The $\alpha$ coefficients are always $\alpha_0 = -1, \alpha_1 = 1$ and are not given

| Steps $k$ | Order $p$ | $\beta_0$ | $\beta_1$ | $\beta_2$ | $\beta_3$ | $\beta_4$ | $\beta_5$ |
|---|---|---|---|---|---|---|---|
| 1 | 1 | $0$ | $1$ | | | | |
| 1 | 2 | $\dfrac{1}{2}$ | $\dfrac{1}{2}$ | | | | |
| 2 | 3 | $-\dfrac{1}{12}$ | $\dfrac{8}{12}$ | $\dfrac{5}{12}$ | | | |
| 3 | 4 | $\dfrac{1}{24}$ | $-\dfrac{5}{24}$ | $\dfrac{19}{24}$ | $\dfrac{9}{24}$ | | |
| 4 | 5 | $-\dfrac{19}{720}$ | $\dfrac{106}{720}$ | $-\dfrac{264}{720}$ | $\dfrac{646}{720}$ | $\dfrac{251}{720}$ | |

**Table A.5** Coefficients of the first few backward differentiation formulas $\sum_{j=0}^{k} \alpha_{k-j} y_{n+1-j} = h\beta_k f_{n+1}$. Following the suggestion of [14] we scale these equations by dividing each one by $\beta_k$, so that the error constant for the $k$-step formula is $-1/(k+1)$

| Steps $k$ | Order $p$ | $\alpha_0$ | $\alpha_1$ | $\alpha_2$ | $\alpha_3$ | $\alpha_4$ | $\alpha_5$ | $\beta_k$ |
|---|---|---|---|---|---|---|---|---|
| 1 | 1 | $-1$ | $1$ | | | | | $1$ |
| 2 | 2 | $\dfrac{1}{3}$ | $-\dfrac{4}{3}$ | $1$ | | | | $\dfrac{2}{3}$ |
| 3 | 3 | $-\dfrac{2}{11}$ | $\dfrac{9}{11}$ | $-\dfrac{18}{11}$ | $1$ | | | $\dfrac{6}{11}$ |
| 4 | 4 | $\dfrac{3}{25}$ | $-\dfrac{16}{25}$ | $\dfrac{36}{25}$ | $-\dfrac{48}{25}$ | $1$ | | $\dfrac{12}{25}$ |
| 5 | 5 | $-\dfrac{12}{137}$ | $\dfrac{75}{137}$ | $-\dfrac{200}{137}$ | $\dfrac{300}{137}$ | $-\dfrac{300}{137}$ | $1$ | $\dfrac{60}{137}$ |

**Table A.6** Coefficients of MEBDF, $\sum_{j=0}^{k} \alpha_{k-j} y_{n+1-j} = h(\beta_0 f_{n+1} + \beta_1 f_{n+2})$

| Steps $k$ | Order $p$ | $\beta_0$ | $\beta_1$ | $\alpha_0$ | $\alpha_1$ | $\alpha_2$ | $\alpha_3$ |
|---|---|---|---|---|---|---|---|
| 1 | 2 | $-1/2$ | $3/2$ | $-1$ | $1$ | | |
| 2 | 3 | $22/23$ | $-4/23$ | $5/23$ | $-28/23$ | $1$ | |
| 3 | 4 | $150/197$ | $-18/197$ | $-17/197$ | $99/197$ | $-279/197$ | $1$ |

## A.3 Implemented Integration Methods for Solving Initial Value Problems in R

In Tables A.7–A.8 we list the codes in the R packages **deSolve** and **deTestSet**.

**Table A.7** Original codes for the methods in the R packages **deSolve**, and **deTestSet**

| R package | R function | Code | Specification | References |
|---|---|---|---|---|
| **deSolve** | lsode | LSODE | BDF, Adams, banded and full Jacobian | [15] |
| **deSolve** | lsodes | LSODES | BDF, Adams, sparse Jacobian | [15] |
| **deSolve** | lsoda | LSODA | Similar to lsode, automatic switch stiff/non-stiff | [17] |
| **deSolve** | vode | DVODE | Similar to lsode, but variable-coefficient method | [4] |
| **deSolve** | daspk | DASPK2.0 | BDF | [3] |
| **deSolve** | radau | RADAU5 | Implicit Runge-Kutta | [13] |
| **deSolve** | rkMethod | new | Explicit Runge-Kutta | [23] |
| **deSolve** | ode.1D, ode.2D, ode.3D | new | Method-of-lines, based on lsodes | [23] |
| **deTestSet** | dopri5 | DOPRI5 | Explicit Runge-Kutta | [14] |
| **deTestSet** | dopri853 | DOPRI853 | Explicit Runge-Kutta | [14] |
| **deTestSet** | cashkarp | new | Explicit Runge-Kutta | [7] |
| **deTestSet** | mebdfi | MEBDFI | MEBDF | [6] |
| **deTestSet** | gamd | GAMD | Generalised Adams method | [16] |
| **deTestSet** | bimd | BIMD | Blended implicit method | [5] |

**Table A.8** Features of the IVP solvers in the R packages **deSolve** and **deTestSet** (See [21])

| Solver | $y' = f(t,y)$ | $My' = f(t,y)$ | $F(y',t,y) = 0$ | Roots | Events | Delays | Compiled functions |
|---|---|---|---|---|---|---|---|
| lsoda,lsodar | $\checkmark$ | | | $\checkmark$ | $\checkmark^*$ | $\checkmark^*$ | $\checkmark^*$ |
| lsode | $\checkmark$ | | | $\checkmark^*$ | $\checkmark^*$ | $\checkmark^*$ | $\checkmark^*$ |
| lsodes | $\checkmark$ | | | $\checkmark^*$ | $\checkmark^*$ | $\checkmark^*$ | $\checkmark^*$ |
| vode | $\checkmark$ | | | | $\checkmark^*$ | $\checkmark^*$ | $\checkmark^*$ |
| daspk | $\checkmark^*$ | $\checkmark^*$ | $\checkmark$ | | $\checkmark^*$ | $\checkmark^*$ | $\checkmark^*$ |
| radau | $\checkmark$ | $\checkmark$ | | $\checkmark^*$ | $\checkmark^*$ | $\checkmark^*$ | $\checkmark^*$ |
| explicit R-K | $\checkmark^*$ | | | | $\checkmark^*$ | | $\checkmark^*$ |
| dopri5 | $\checkmark$ | | | | | | $\checkmark^*$ |
| dopri853 | $\checkmark$ | | | | | | $\checkmark^*$ |
| cashkarp | $\checkmark$ | | | | | | $\checkmark^*$ |
| mebdfi | $\checkmark^*$ | $\checkmark^*$ | $\checkmark$ | | | | $\checkmark^*$ |
| gamd | $\checkmark^*$ | $\checkmark$ | | | | | $\checkmark^*$ |
| bimd | $\checkmark^*$ | $\checkmark$ | | | | | $\checkmark^*$ |

"Compiled functions" means that the differential equations can be specified in compiled code
$\checkmark$ denotes that the feature was present in the original code, $\checkmark^*$ means that it was added in the R implementation

## A.4   Other Integration Methods in R

In Table A.9 we list the codes in the R packages **bvpSolve** and **rootSolve**.

**Table A.9** Original codes and references of the methods in the R packages **bvpSolve** (BVP), and **rootSolve** (BVP, PDE)

| R package | R function | Original code | Specification | References |
|-----------|------------|---------------|---------------|------------|
| **bvpSolve** | bvpshoot | new | Single shooting | [22] |
| **bvpSolve** | bvpcol | COLNEW | Collocation, monomial spline | [2] |
| **bvpSolve** | bvpcol(bspline = TRUE,...) | COLSYS | Collocation, bspline | [1] |
| **bvpSolve** | bvpcol(epsini = ...) | COLMOD | Collocation, automatic continuation | [11] |
| **bvpSolve** | bvptwp | TWPBVP | MIRK | [10] |
| **bvpSolve** | bvptwp(cond = TRUE,...) | TWPBVPC | MIRK with conditioning | [8] |
| **bvpSolve** | bvptwp(lobatto = TRUE,...) | TWPBVPL | MIRK based on lobatto | [9] |
| **bvpSolve** | bvptwp(epsini = ...) | ACDC | MIRK based on lobatto with automatic continuation | [11] |
| **bvpSolve** | bvptwp(lobatto = TRUE,cond = TRUE, ...) | TWPBVPLC | MIRK based on lobatto with conditioning | [9] |
| **rootSolve** | steady.1D | new | BVP, finite differencing | [20] |
| **rootSolve** | steady.2D, steady.3D | new | PDE, finite differencing, uses Yale Sparse Matrix package and SPARSKIT | [12, 18, 19] |

# References

1. Ascher, U. M., Christiansen, J., & Russell, R. D. (1979). **COLSYS** – a collocation code for boundary value problems. In B. Childs et al. (ed.), *Codes for boundary-value problem, in ordinary differential equations: Vol. 76. Lecture notes computer science* (pp. 164–185). Berlin/Heidelberg/NewYork: Springer.
2. Bader, G., & Ascher, U. M. (1987). A new basis implementation for a mixed order boundary value ODE solver. *SIAM Journal on Scientific and Statistical Computing, 8*, 483–500.
3. Brenan, K. E., Campbell, S. L., & Petzold, L. R. (1996). *Numerical solution of initial-value problems in differential-algebraic equations. SIAM classics in applied mathematics.* Philadelphia, PA: Society for Industrial and Applied Mathematics.
4. Brown, P. N., Byrne, G. D., & Hindmarsh, A. C. (1989). **VODE**, a variable-coefficient ODE solver. *SIAM Journal on Scientific and Statistical Computing, 10*, 1038–1051.
5. Brugnano, L., Magherini, C., & Mugnai, F. (2006). Blended implicit methods for the numerical solution of DAE problems. *Journal of Computational and Applied Mathematics, 189*(1–2), 34–50.

6. Cash, J. R., & Considine, S. (1992). An **MEBDF** code for stiff initial value problems. *ACM Transactions on Mathematical Software, 18*(2), 142–158.
7. Cash, J. R., & Karp, A. H. (1990). A variable order Runge–Kutta method for initial value problems with rapidly varying right-hand sides. *ACM Transactions on Mathematical Software, 16*, 201–222.
8. Cash, J. R., & Mazzia, F. (2005). A new mesh selection algorithm, based on conditioning, for two-point boundary value codes. *Journal of Computational and Applied Mathematics, 184*, 362–381.
9. Cash, J. R., & Mazzia, F. (2006). Hybrid mesh selection algorithms based on conditioning for two-point boundary value problems. *Journal of Numerical Analysis, Industrial and Applied Mathematics, 1*(1), 81–90.
10. Cash, J. R., & Wright, M. H. (1991). A deferred correction method for nonlinear two-point boundary value problems: Implementation and numerical evaluation. *SIAM Journal on Scientific and Statistical Computing, 12*, 971–989.
11. Cash, J. R., Moore, G., & Wright, R. W. (1995). An automatic continuation strategy for the solution of singularly perturbed linear two-point boundary value problems. *Journal of Computational Physics, 122*, 266–279.
12. Eisenstat, S. C., Gursky, M. C., Schultz, M. H., & Sherman, A. H. (1982). Yale sparse matrix package. I. The symmetric codes. *International Journal for Numerical Methods in Engineering, 18*, 1145–1151.
13. Hairer, E., & Wanner, G. (1996). *Solving ordinary differential equations II: Stiff and differential-algebraic problems*. Heidelberg: Springer.
14. Hairer, E., Norsett, S. P., & Wanner, G. (2009). *Solving ordinary differential equations I: Nonstiff problems* (2nd rev. ed.). Heidelberg: Springer.
15. Hindmarsh, A. C. (1980). **LSODE** and **LSODI**, two new initial value ordinary differential equation solvers. *ACM-SIGNUM Newsletter, 15*, 10–11.
16. Iavernaro, F., & Mazzia, F. (1998). Solving ordinary differential equations by generalized Adams methods: Properties and implementation techniques. *Applied Numerical Mathematics, 28*(2–4), 107–126. Eighth conference on the numerical treatment of differential equations (Alexisbad, 1997).
17. Petzold, L. R. (1983). Automatic selection of methods for solving stiff and nonstiff systems of ordinary differential equations. *SIAM Journal on Scientific and Statistical Computing, 4*, 136–148.
18. Saad, Y. (1994). **SPARSKIT***: A basic tool kit for sparse matrix computations. VERSION 2*.
19. Saad, Y. (2003). *Iterative methods for sparse linear systems* (2nd ed.). Philadelphia, PA: Society for Industrial and Applied Mathematics.
20. Soetaert, K. (2011). **rootSolve***: Nonlinear root finding, equilibrium and steady-state analysis of ordinary differential equations*. R package version 1.6.2.
21. Soetaert, K., Petzoldt, T., & Setzer, R. W. (2009). *R-package* **deSolve***, writing code in compiled languages*. Package vignette.
22. Soetaert, K., Cash, J. R., & Mazzia, F. (2011). **bvpSolve***: Solvers for boundary value problems of ordinary differential equations*. R package version 1.2.2.
23. Soetaert, K., Petzoldt, T., & Setzer, R. W. (2010). Solving differential equations in R: Package **deSolve**. *Journal of Statistical Software, 33*(9), 1–25.

# Index